Wood Framing Portable Handbook

Wood Framing Portable Handbook

Jonathan F. Hutchings

McGraw-Hill

New York San Francisco Washington, D.C. Auckland Bogotá
Caracas Lisbon London Madrid Mexico City Milan
Montreal New Delhi San Juan Singapore
Sydney Tokyo Toronto

Library of Congress Cataloging-in-Publication Data

Hutchings, Jonathan
Wood framing portable handbook / Jonathan F. Hutchings
p. cm.
ISBN 0-07-134247-8
1. Framing (Building) Handbook, manuals, etc. I. Title.

TH2301.H88 2000
694.2—dc21 99-32251
CIP

McGraw-Hill

A Division of The ***McGraw-Hill*** *Companies*

1 2 3 4 5 6 7 8 9 0 DOC/DOC 0 9 8 7 6 5 4 3 2 1 0

ISBN 0-07-134247-8

The sponsoring editor for this book was Zoe Foundotos, the editing supervisor was Suzanne Ingrao, and the production supervisor was Pamela A. Pelton. It was set in Century Schoolbook per the MHT 5 x 8 design by Michele Pridmore and Paul Scozzari of McGraw-Hill's Hightstown, N.J., Professional Book Group composition unit.

Printed and bound by R. R. Donnelley & Sons Company.

To the pride and respect which is the hallmark of every good American carpenter.

We're the best the world has ever known.

Copyright acknowledgements to Simpson Strong-Tie Company for information, data, and illustrations on modern structural framing connectors.

Contents

Foreword

Sophistication in structural framing techniques is rapidly becoming one of the bottom line necessities of business longevity for the American building contractor. Achieving a profit consistently in this business means using competitive knowledge advantageously in all phases of project operations and especially the big ticket items like framing.

Next-step technology has brought the American builder such things as noncombustible, structural, fiber-reinforced cement board; oriented strand board (OSB); panelized shear wall; and structural floor and roof trusses that are lighter and stronger than traditional solid sawn wood members. Building codes are getting tougher all the time in structural wood and framing code requirements as well as the new related code-required earthquake seismic connectors.

This book is unique in its comprehensive coverage as an illustrated pocket guide for residential framing. Every residential contractor needs this handbook onsite every day in his or her nailbags or toolbox—the structural framing and engineered wood products chapters alone are worth it. I've known the author of this book for 20 years and he knows as well as I how much the American construction industry has changed in these last two decades. Cost over-runs cannot be absorbed by the contractor any more because the tight profit margin on the bid will simply not allow absorption of any extra cost without direct reciprocal profit loss.

Cost-effective structural framing techniques are looking to next-step construction technology for project cost savings. The modern construction industry is so competitive one must be aggressive in the arena of applying those techniques daily. Over the past decades this industry has seen a sharp increase in cost of labor, labor burden costs, overhead, and time constraints on project execution. Paralleling this trend has been an increase in the amount of cost in construction and materials. This has resulted in a profit squeeze for today's builder dependent on bringing the project in on budget in order to take a paycheck home.

This portable framing handbook will definitely be on the jobsite with my framing crews and if you're serious about carving out a successful career in framing construction for yourself, my advice to you is to grab this very special book in a hurry before they sell out.

Chad Simmons, President
SJB Construction Corporation
Silicon Valley, California

Preface

At this time on this planet, Western Platform Framing is the accepted standard of excellence in home-building and light commercial construction. Our structural framing techniques work so well that many countries send their building professionals to us to learn platform framing. After serving as an engineer for the insurance industry's reconstruction efforts in both the 7.6 Loma Prieta earthquake of 1989 and later in the 7.2 Northridge-Los Angeles earthquake of 1993, and having witnessed the devastation firsthand, I understand why the rest of the world sends their building officials to America to learn our ways of structural framing. When one stops and reflects, the reason the world looks to us for leadership in framing construction is quite clear:

We build the best. Western Platform Framing produces the safest, strongest homes in the history of humankind.

That is not just my opinion, that is documented fact. The statistics bear me out: 61 casualties in the Northridge quake compared to 5000 in Kobe, Japan. Both quakes registered 7 plus on the Richter scale, with Northridge shaking a little harder. Same relative seismic activity, but 61 deaths compared to *5000*. The difference (in part) is due to our highly evolved techniques and standards of structural framing. American builders are the last proud American industry still retaining global leadership, and justifiably so.

Now, if you're a builder with some years of experience under your belt, that last paragraph brought some of the pride back to your heart, didn't it? That's because craftsmanship is a matter of respect which should transcend the money. Down at the bottom, that's what the American builder is all about—the fire-, life-, safety-protection of the people who will inhabit the builder's structure. Houses don't forget. You will walk away from the job one last time and then others will live there long after you've gone, but it will always be *yours*. If the house came to exist and serve humankind from the sweat of your hands and the ache of your muscles, then you will always be a part of it—even years down the line after they've shoveled dirt on your coffin. It will always be yours. Houses don't forget.

Today's American builders, like business people everywhere who work too hard, are always reassessing their production costs. This book is written with one purpose in mind: to provide the builder with an illustrated practical application knowledge of the various framing systems and techniques used in modern residential construction to increase that bottom line production profit.

Structural framing is designed to be supported by a perimeter stemwall foundation in addition to individual supports or piers and anchored with a number of framing-to-foundation tiedowns and framing-to-roof anchoring system holdowns. These are collectively referred to as framing anchoring systems, and this book covers the most widely used methods of these framing systems and accepted standards for anchoring the structural framing to the foundation and to the roof.

Any structural framing system must resist vertical loads from the weight of the framing, plus temporary extra roof loading, and it must resist side loads imposed on the structure by wind blowing against the walls. Design data describing the roof and wind loads which the home has been designed to resist, must be called out in the original design framing plans. Building codes load zone maps of the United States showing roof load, wind load, and thermal zones are also used in these calculations. Anchoring systems are then used to fasten the structural framing to the foundation and each load beam or pier at stemwall grade level.

These modern frame-to-foundation and frame-to-roof connections are rated for resistance to uplift forces used as factors in Zone I and Zone II wind areas along the Gulf of

Mexico and East Coast, and earthquake Seismic Zone 4 requirements along the West Coast. Individual codes governing framing anchoring systems in your area may have specific wind and roof load design requirements which vary from the general information in this handbook so when in doubt, check with your local building official.

Remember, in matters of real-world applications of your construction business, always research the plans and specifications thoroughly for code compliance and consult an attorney in all contractual obligation matters. This book does not offer legal advice, nor should it be construed as such. Further, please be informed that the facts, subjects, references, and standards contained herein are as accurate as this time of writing will allow.

Respectfully submitted,
Jonathan F. Hutchings

Acknowledgments

The following individuals and organizations contributed to the making of this book. I would like to take this brief moment to thank them and give them their deserved moment of attention in the spotlight. Both of these organizations are industry leaders and I am very grateful for their expert help in compiling this Work.

Richard R. Chapman, Simpson Strong-Tie Company, Pleasanton, California

William B. Ware, General Counsel, Southern Building Code Congress International, Birmingham, Alabama

Written permission to reproduce this material was sought from and granted by the copyright holder, Southern Building Code Congress International, 900 Montclair Road, Birmingham, Alabama 35213-1206.

Chapter

1

General

Wood Light Frame Construction

Wood light frame construction is the most versatile and flexible of all building systems. It can be used to provide a structural frame for any shape from a plain rectangle to complex cylindrical towers, in any size building from single to multistory. In American construction, there are three types of wood light frame construction: platform framing (also called western platform framing), balloon framing, and post-and-beam framing.

Platform framing is the most modern and common type of framing. In this type of framing, the subfloor of the first story extends to the outside edges of the foundation and forms a platform upon which the walls of the house rest. Each story is built upon its own platform so settlement and shrinkage throughout the structural framing is displaced uniformly. We'll look at platform framing extensively in a moment after the following section on balloon framing.

Balloon Framing

Studs in balloon framing rest on the sill and extend unbroken from the foundation to the roof (see Fig. 1-1). Balloon frames using full-length studs running two stories from foundation to roof were eventually recognized as being too long to erect efficiently, and the hollow spaces between the studs (stud cavities) acted as multiple chimneys in a fire, spreading the blaze rapidly to the upper floors unless closed off with firestops at each floorline.

Firestops were then required at all floor levels in balloon framing to close the stud cavities in the wall against the upward passage of fire. Floor joists in balloon framing rest on a wood ribbon (also called a ribband) recessed into the studs. In platform framing, the studs and sole plates rest on the top of the floor platform (subfloor).

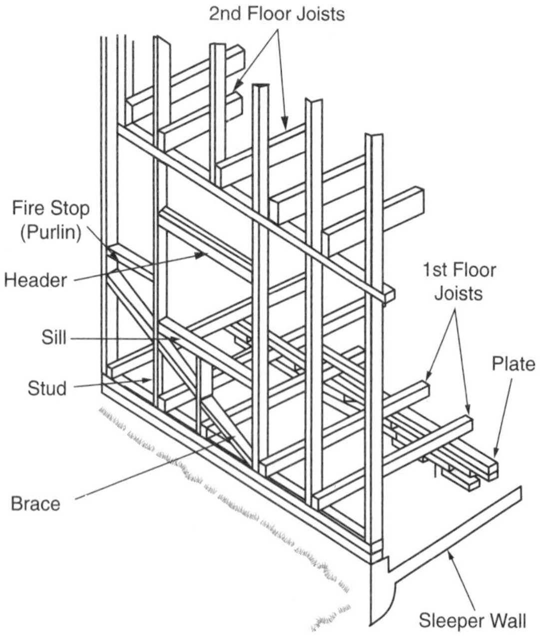

Figure 1-1 Balloon Framing

Balloon framing does have its advantage over platform framing because the platform frame will shrink and contract over time and cause considerable settlement of the structural frame. Construction-grade framing lumber is relatively fresh from the mill these days and only marginally dried. We in the business call this "pond-dried lumber" (dried at the bottom of a pond).

This framing lumber shrinkage is part of the business of constructing light wood platform frame structures and tends to cause drywall and plaster cracking as the house settles and ages. If the framing wood is terribly wet, I've seen windows get cracked from the resulting shrinkage.

Platform Framing

Wood light framing in the form of platform framing is popular in the industry because it is an extremely flexible and economical way of constructing small buildings. Its flexibility stems from the ease with which carpenters with ordinary tools can create buildings of astonishing complexity in a variety of geometrics. Its economy can be attributed in part to the relatively unprocessed nature of the materials from which it is made, and in part to mass-market competition among suppliers of components and materials and local competition among small builders.

Platform framing (Fig. 1-2) is the one truly complete and open system of construction we have. It incorporates structure, enclosure, thermal insulation, mechanical installations, and finishes into a single construction without scaffolding, and overhangs can be eliminated to save material. Thousands of products are made to fit it: dozens of competing brands of windows and doors; a multitude of interior and exterior finish materials; scores of electrical, plumbing, and heating products. For better or worse, it can be dressed up to look like a building of wood or of masonry in any architectural style from any era of history. Architects have failed to exhaust its formal possibilities and engineers have failed to invent a new environmental control system it cannot assimilate.

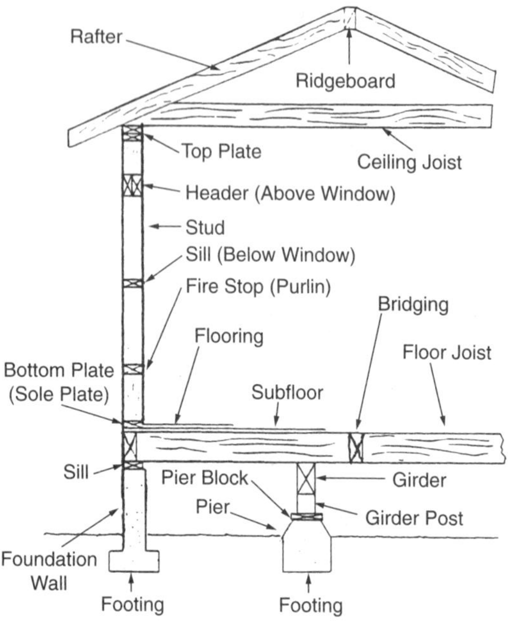

Figure 1-2 Platform Framing

At the other end of the scale, cheap pop-outs of platform frame construction have led to the creation of countless ugly cracker boxes around the perimeters of our cities. To create the minimum-cost, detached, single-family house, one can begin by reducing the number of exterior wall corners to four because extra corners cost extra dollars for framing and finishing. Then the roof pitch can be made as shallow as asphalt shingles will permit, to allow roofers to work and labor. Windows can be made as small as building codes will allow, because a square foot of window costs more than a square foot of wall. The thinnest, cheapest finish materials can be used inside and out, even if they will begin to look shabby after a short period of time, and all ceilings can be made flat and as low as is legally permitted, to save materials and labor. The result is an uninteresting, squat, dark residence that weathers poorly, has little meaningful relation to its site, and contributes to a lifeless, uninspiring neighborhood. The same principles, with roughly the same result, can be applied to the construction of apartments and commercial buildings. In most cases, things only get worse if one tries to dress up this minimal box building with inexpensive shutters, window boxes, and fake materials.

The single most devastating fault of wood light frame construction is not its combustibility or its tendency to decay, but its ability to be reduced to the minimal—scarcely human—building system. Yet one can look to examples of the Carpenter Gothic, Queen Anne, and Shingle Style buildings of the century past, or the Bay Region and Modern styles of our own time, to realize that wood light framing can be a system that gives the architect the freedom to make a finely crafted building that nurtures life and elevates the spirit.

While complex in its details (Fig. 1-3), the platform frame is very simple in concept: A floor platform is built. Loadbearing walls are erected upon it. A second floor platform is built upon these walls, and a second set of walls upon this platform. The attic and roof are then built upon the second set of walls. There are, of course, many variations: A concrete slab is often substituted for the ground floor platform; a building may be one or three stories tall instead of two; and several types of roofs are frequently built that do not incorporate attics. But the essentials remain: A floor platform is completed at each level, and the walls bear upon the platform rather than directly upon the walls below.

The advantages of the platform frame over the balloon frame are that it uses short, easily handled lengths of lumber for the wall framing; its hollow spaces are automatically firestopped at each floor, and its platforms are convenient working surfaces for the carpenters who build the frame. The major disadvantage of the platform frame is that each platform constitutes a thick layer of wood whose grain runs horizontally. This leads inevitably to a relatively large amount of vertical shrinkage in the frame as the excess moisture dries from the wood, which can lead to distress in the exterior and interior finish surfaces, especially around multistory spaces such as stairwells and tall rooms.

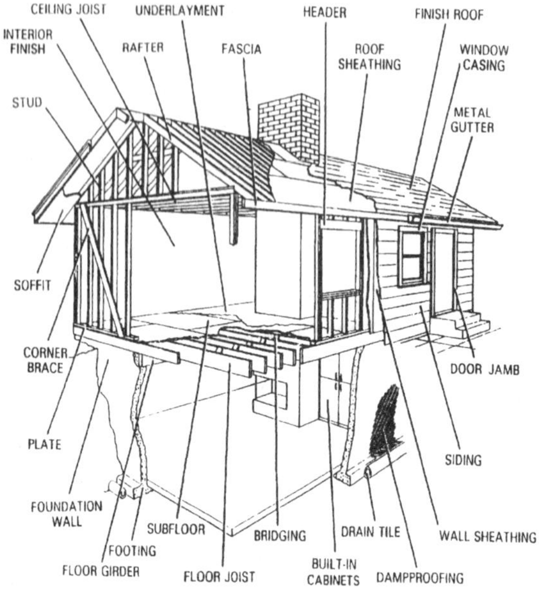

Figure 1-3 Platform Framing Details

A platform frame is made entirely of nominal 2-inch (in) members, actually 1 $^1/_2$ in (38 mm) in nominal thickness. These are ordered and delivered cut to the nearest 2-ft (600-mm) length and then are measured and sawn to exact length on the building site. Most framing connections are made with nails, using either face nailing, end nailing, or toenailing as required by the characteristics of each joint. Nails are driven either by hammer or by pneumatic nailing machines. In either case, the connection is quickly made because the nails penetrate the wood without the need for drilling holes or otherwise preparing the joint.

Each plane of structure in a platform frame is made by aligning a number of pieces of framing lumber parallel to one another at specific intervals, nailing these to crosspieces that *head off* the framing at either end to maintain its spacing and flatness, then covering the plane of framing with *sheathing* of boards or panels that join and stabilize the pieces into a single structural unit, ready for the application of finish materials inside and out. In a floor structure, the parallel pieces are the floor joists, and the crosspieces at the ends of the joists are called *headers*, *rim joists*, or *band joists*. The sheathing on a floor is known as the *subfloor*. In a wall structure, the parallel pieces are the studs, the crosspiece at the bottom of the wall is the *sole plate*, and the crosspiece at the top (which is doubled for strength if the wall bears a load from above) is called the *top plate*. In a sloping roof, the rafters are headed off by the top plates at the lower edge of the roof, and by the *ridge board* at the peak.

Openings are required in all these planes of structure: for windows and doors in the walls; for stairs and chimneys in the floors; and for chimneys, skylights, and dormers in the roofs (Fig. 1-4). In each case, these are made by heading off the opening as follows:

- Openings in floors are framed with *headers* and *trimmers*
- In walls, *sills* head off the bottoms of openings, with strong *window headers* and *door headers* across the tops
- Headers and trimmers must be doubled or otherwise strengthened to support the higher loads placed on them by the presence of the opening

Sheathing is the component of platform framing that enables the concept to work. The end nails that connect the plates to the studs have little holding power against uplift of the roof by wind, but the sheathing connects the frame into a single, strong unit from foundation to roof. The rectilinear geometry of the parallel framing members has no useful resistance to *wracking* by lateral forces such as wind, but rigid sheathing panels or diagonal sheathing boards brace the building effectively against these forces. Sheathing also furnishes a surface to which shingles, boards, and flooring are nailed for finish surfaces. In buildings without sheathing, or with sheathing materials that are too weak to tie and brace the frame, diagonal bracing must be used for lateral stability.

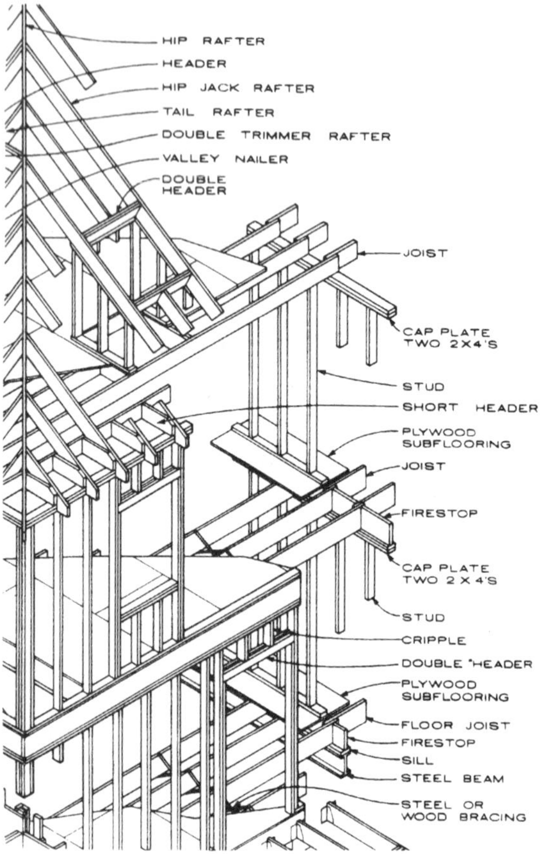

Figure 1-4 Framing Openings

The platform framing system is suitable for both one-story and multistory structures, and most modern residential construction has a structural frame of platform framing. The system is used on both perimeter and slab-on-grade jobs. The framing consists mainly of construction grade Douglas Fir (or Southern Pine) wood in nominal sized 2 × 4s, 2 × 6s, and wider boards that may be used as studs, joists, and plates. A house with platform framing is like a stack of lidless boxes topped with a roof. During construction, the walls—studs, sills, and plates—are nailed together on the subfloor and raised into place. The economic advantages of platform framing are:

- The system uses short, easily-handled lengths of lumber for the wall framing
- The stud cavities are automatically firestopped at each floor
- Its platforms are convenient working surfaces for the carpenters who build the frame

The major disadvantage of platform framing is that each platform constitutes a thick layer of wood whose grain runs horizontally. This leads to the relatively large amount of vertical shrinkage mentioned earlier. As the excess moisture dries from the wood, it causes distress in the exterior and interior wall finish surfaces, especially around large open-framed areas such as stairways and cathedral or vaulted ceilings.

Many houses and apartment buildings built prior to the end of World War II have balloon framing. In this framing system, the walls are nailed directly to the studs (generally 2 × 4s) that extend the entire height of the frame, from the sill to the roof plate, even when there is more than one story. In balloon framing, studs form a seamless ribbed case, like the frame of a squared-off hot-air balloon (hence the name).

In this system of framing, studs are nailed to an anchored sill and floor joists are nailed to the sides of studs. Wood firestops of the same thickness and width as the studs should be present at each floor level, at the midpoints between the floors, and between the outer ends of the joists. These keep the vertical and horizontal spaces from acting as flues in the event of fire. Balloon framing is rarely used anymore because of the high cost and general unavailability of the long lumber needed for the studs. Balloon framing also does not perform as well against seismic forces in an earthquake.

An even older type of framing (post-and-beam framing), however, is enjoying a revival because of the expansive living spaces it creates inside a home. Horizontal wood beams, which support the subfloors and roof of the house, are connected to upright wood posts with mortise-and-tenon joints or some other strong joint. A few posts take the place of many studs, and joists are notched into the beams.

Studs can be added between posts for framing doors and windows but are not part of the supporting structure. The beams are generally 6 in wide and 10 in deep. Posts that support the subfloors are generally 8 in thick; those that hold up the roof, 6 in thick. The posts are set 6, 7, or 8 ft apart, in contrast to other wood-framing techniques in which studs are set 16 or 24 in apart. This type of structural framing gives a strong rustic timber look to the architecture of the building.

Framing Fasteners

Before the development of modern timber connectors and glued-laminated members, lumber was joined together by framing with interlocking cut joints. These joints were weak and required lumber large enough so that the uncut part of the framing member would hold the required loads.

In recent years, the kinds of joints and their placement have changed dramatically. The emphasis in framing is now on better design and making material of smaller dimensions do more work without losing necessary strength. The development of new stronger types of fasteners has made this new emphasis on design possible.

However, this does not mean that all the old construction ways were bad. In fact, the more we learn about the traditional framing methods, the more we realize that a surprising amount of "new" practices are really adaptations of older and fundamentally sound ways of building. For example, the traditional method of framing involved cutting lumber to form joints held together with various types of dowels, wedges, and pins. Later, cut nails made from metal plates came into use as well as bolts and metal straps to reinforce and supplement the use of joints.

With the development of modern nail-making machinery and the production of other types of fasteners, the use of many of the traditional framing joints has been largely discontinued. Now other more simple joints are used, permitting the use of lighter materials for longer spans which have equal or greater loadbearing capacities.

Nails

Nails and spikes are the most common mechanical fasteners used in the construction industry (Fig 1-5). They come in many different sizes, head and point styles, shank types, kinds of metal, and finishes. The sizes of standard nails are indicated by an old system that uses the abbreviation "d" for *penny*. The shortest nail in this system is the 2d (two penny), which is 1 in long. The longest is the 60d (60 penny), which is 6 in long. Nails shorter than the 2d are indicated by length and gauge number.

The common wire nail is made of steel; however, other metals and alloys are used, including brass, copper, stainless steel, and aluminum. The use of these other materials is indicated when there is a need for resistance to stain or corrosion caused by exposure to weather, acids, alkaline

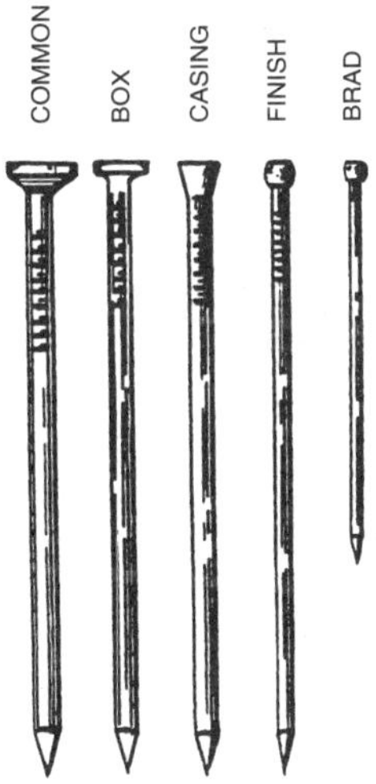

Figure 1-5 Framing Nails

solutions, brine, or fumes. When nails require corrosion resistance and metals other than steel lack sufficient strength, steel nails may also be given certain surface treatments. This surface treatment may also be used to improve the appearance of the nail or for purposes of sanitation.

Some nails are also hardened to permit driving into concrete or masonry, while others are annealed or softened to permit riveting. The surface of the nail may be roughened or the shank may be deformed to increase holding power.

Splitting may occur in the nailing of any of the denser hardwoods and some of the commonly used softwood framing lumber such as Douglas Fir, Southern Pine, and Hemlock. Methods of preventing splitting include the use of lighter-gauge nails, the boring of pilot holes, and clamping or applying pressure where the nail is driven. In hardwoods, the lighter-gauge nails may bend in driving so pilot holes may be necessary under any circumstances. Blunting the point of the nail may also help to prevent splitting.

This is often done in finish work, where less holding power is required, but is not recommended in framing where maximum holding power is essential. Besides the shape of the point and shank, factors which affect the holding power of nails include direction of driving, clinching, seasoning of the lumber, and surface finish of the nail.

Nails driven perpendicular into the grain of the wood have greater holding power than nails driven parallel to the grain or into the endwood. Nails driven at a slant with the surface (toenailing), rather than perpendicular to it, generally have greater holding power. If they are cross slanted such that some are slanted one way and an equal number slanted the opposite way, the holding power is increased still more.

Toenailing makes a stronger joint than nailing through one piece perpendicular to the grain into the end grain of another. Clinching, or the bending over of the protruding part, increases holding power. This is greater if the clinching is done across the grain, although it does not present as neat an appearance. Nails driven into seasoned lumber have greater holding power than if driven into the same species of green lumber. This is still true even if the green lumber is seasoned after the nail is driven.

The gang nail is a prefabricated truss connector plate. It is fabricated from galvanized steel of various gauges. The shanks of the nail are pressed out of the plate. The advantage of the gang nail is speed in production of roof trusses. The chemically etched nail developed by the Forest Products Laboratory has greater holding power than the plain nail by about 180 to 200 percent in softwoods and from 90 to 175 percent in hardwoods. The sandblasted nail is comparable. Under impact loading, it has roughly the same holding power as the cement coated or plain nail. In softwood, the cement coated nail has from 85 to 100 percent increased withdrawal resistance but has little advantage in hardwood.

Nails are the preferred fasteners in residential framing for several reasons (see Fig. 1-6). They cost less to install than screws and are easier to install in volume due to pneumatic tools commonly known as "nail guns." Engineers prefer nails because they are generally more ductile than screws, and better ductility equates to better absorption of seismic energy. When screw withdrawal properties are desirable, such as floor sheathing, ring-shank or screw-shank nails can be used instead.

Most structural designers specify common nails because of their improved strength and stiffness over box, cooler, or sinker nails. Common nails also have larger nail shank diameters than other nails. Although hot-dip galvanized box nails may be used for plywood sidings, common nails are recommended to fasten wood structural panel shear walls.

Hand-driven nails typically come in boxes labeled with the nail type. Boxes with nails for nail guns sometimes show only the length, diameter, and finish. To verify the use of common nails, the framer must be familiar with nail diameter requirements. For example, if a nail gun box says 2 $^3/_8$ × .113 Smooth instead of 2 $^3/_8$ × .131 Smooth, the fastener is an 8d cooler nail instead of the 8d common. This is a common framing construction error. Fig. 1-7 shows a box label for common nails, and Fig. 1-8 shows a performance comparison of common nails versus box nails.

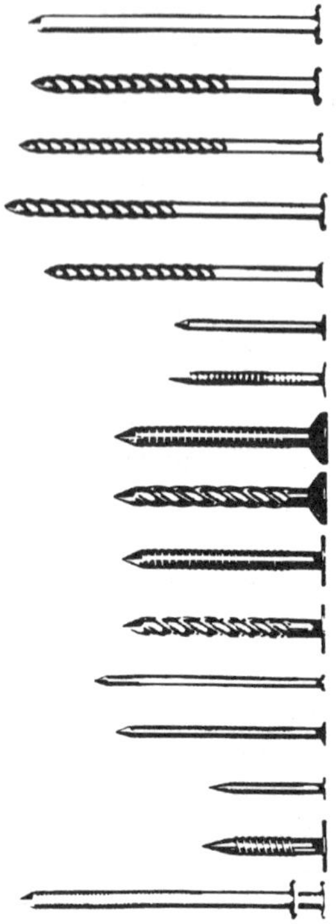

Figure 1-6 Nail Types

Shortened 10d common nails, commonly referred to as *short* or *plywood nails*, come in three lengths: 2$^{1}/_{8}$, 2$^{1}/_{4}$, and 2$^{3}/_{8}$ in. Only $^{11}/_{32}$-in-thick panels can use the shortest nail, 2 $^{1}/_{8}$ in. Panels with thicknesses from $^{3}/_{8}$ in to $^{15}/_{32}$ in can use the 2$^{1}/_{4}$-in length nail. Panel thicknesses from $^{1}/_{2}$ in to $^{19}/_{32}$ in require the 2 $^{3}/_{8}$-in length nail. Panel thicknesses equal to $^{5}/_{8}$ in or thicker must use the full length 10d common (3 in).

Unless otherwise specified, shortened 10d common nails may fasten wood structural panels in shear walls if they meet the minimum penetration requirements. All nails must provide a minimum of 12 diameters penetration into the framing member or blocking. Because of their shallower penetration, shorter nails are less likely to split framing members but are more likely to withdraw. Check with the

Figure 1-7 Box Label for Common Nails

COMPARISON OF COMMON VS. BOX NAILS								
Nail Size	Length, inches	Shank Diameter, decimal inch		Minimum Spacing and Penetration, inches		Allowable Lateral Load, pounds		
		Common Nail	Box Nail	Common Nail	Box Nail	Common Nail	Box Nail	Box vs. Common
6d	2	0.113	0.099	1 1/4	1 1/8	63	51	0.81
8d	21/2	0.131	0.113	1 1/2	1 1/4	78	63	0.81
10d	3	0.148	0.128	1 5/8	1 1/2	94	76	0.81

Figure 1-8 Common vs. Box Nails

structural designer or building official before using these nails.

Because the head sizes of common and box nails (Fig. 1-9) are nearly identical, wood structural panel sheathing fastened with these nails will reach the same ultimate strength. The advantage of using common nails to fasten sheathing is generally stiffness and not strength. Common nails have less slip than box nails due to their increased shank diameter. Because shear walls must provide both strength and stiffness, the use of box, cooler, or sinker nails will reduce the performance of the shear wall.

For sheathing, four common nails provide the same stiffness as five box or cooler nails. When shear walls are constructed with box, sinker, or cooler nails, it will take 25 percent more nails to make up the difference. However,

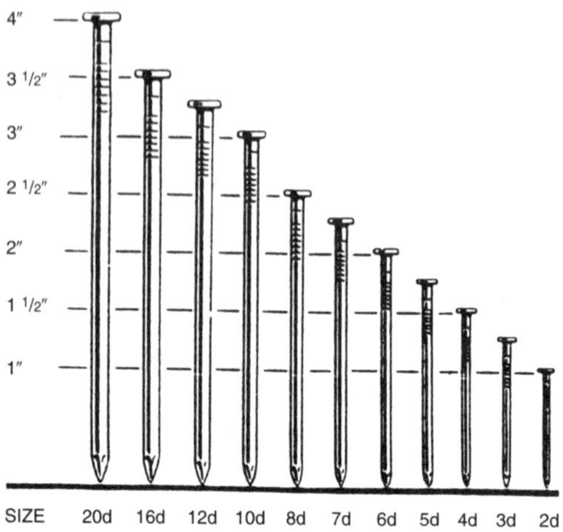

Figure 1-9 Nail Sizes

additional nailing must comply with the following requirements for minimum spacing and framing member sizes:

- When additional nailing creates spacing of 6d or 8d at 2 in on center or 10d at 3 in on center, 3-in nominal width framing members are required at all adjoining panel edges.
- When the common nail spacing is equal or less than 6d at 2 in, 8d at 2 ½ in, or 10d at 3 in, additional nails cannot be used because of minimum spacing requirements. In these cases, additional sheathing must be provided and is usually installed on the other side of the same wall.

Most framers use nail guns to fasten wood structural panel sheathing. The nails used in these tools come in clips that may or may not have full heads. Some manufacturers clip the nail heads to form clips with the nail shanks side by side.

This arrangement allows more nails per clip, reduces the frequency of reloading, and avoids intershank plastic from jamming and disabling the tool. Other manufacturers use plastic to hold the individual nails far enough apart to allow full heads.

Although present building codes accept altered heads, the codes are being changed because there are good reasons to believe altered heads are not as strong as full-headed nails. Examination of plywood shear walls damaged in the Northridge—Los Angeles earthquake showed the importance of nail head size.

Through investigation of the localized destruction of that earthquake, it was determined that shear wall failure occurred at the panel edges; the panel edge itself either failed by punching through at the nail head or withdrawal of the nail itself in panel buckling. In both cases, the nail head size was a significant element in the mode of failure: The smaller the nail head sizes, the more easily a panel can buckle off the framing.

Tests that established the strengths of wood structural panels used hand-driven nails with full heads. Some tests used casing nails, and these tests showed results that shear walls with 8d casing nails are only 62 percent as strong as ones with 8d box nails. This is due to the fact that although the shank diameters of the two nails are the same, the head area of the casing nail is only 27 percent that of the box nail.

Spikes

Spikes are made in two general classes. One includes those made of wire, like nails, while the other includes spikes made from square bars. The latter are commonly referred to as boat or ship spikes; the common railroad track spike is in this class.

Common wire spikes vary from 3 in (10d) to 12 in long, with gauges from No. 6 to $^3/_8$ in. Although their lengths coincide with that of standard wire nails, their thicknesses are greater. *Cut spikes* vary in length from $3^1/_2$ to 7 in. The lengths of *boat spikes* vary from 3 to 16 in, while the thickness varies from $^1/_4$ to $^5/_8$ in. The *track spike* is similar except that it has a head that projects on one side for holding the base of the rail. It is 6 in long and $^5/_8$ in thick in the largest size.

Drift Bolts

Drift bolts, sometimes called *pins*, are simply lengths of round or square steel rods which are driven like spikes. They may have heads at one end, but they are not threaded. They require holes bored into the wood where they are to be driven. These holes must be smaller than the diameter of the drift bolt to produce gripping friction.

The actual size of the hole may vary, according to the kind of wood, the size of the bolt, and the depth of penetration. In general, a hole bored to a size of about $^2/_3$ to $^4/_5$ of the diameter of the bolt seems to develop the greatest holding power. Drift bolts find much use in timber framing and they are also the most economical fastener for temporary framing work.

Threaded Fasteners

Metal straps and threaded fasteners may be used to advantage where members meet at any angle. Some are ready-made for standard conditions; others may be made up in the shop or field to fit specific needs. They may be used with threaded fasteners or combinations of two or more fasteners. A stra~p may be used with a bolt welded to one or both ends. A checklist of these fasteners is as follows:

Tie Rods and Cables. Tie rods are very long bolts which hold the lower ends of member trusses from spreading apart under load. They may be used in any situation where an especially strong tension member is needed. Steel cables may be used around the base of domes to hold the arches from spreading, or around staves or lagging. They are also used as tension supports in arches and trusses, or as guys to steady any type of high structure.

Wood Screws. The most important use of screws in the building industry is in fastening other materials to wood, particularly building hardware. Trim is often installed with screws, so that it can be subsequentially removed without damaging any of the parts. Screws, if properly installed, present a neat appearance as well as superior holding power. Two disadvantages are their increased cost over nails and the fact that they require more time to drive.

Common wood screws come in three regular style heads. These are the flat, oval, and round heads. The drive screw and Phillips head screw have flat heads. The length of the flat head screw is its total length, including the head. The length of the round head screw is taken from its point to the junction of the body and the head, while the length of the oval is taken from the point to the edge of the head at its greatest diameter.

Wood screws are stocked in a great many different lengths and thicknesses, ranging from $^1/_4$ to 5 in in length and in gauge from 0 to 24. The larger the gauge number, the larger is the diameter of the screw, which is the opposite of the gauge numbering system for wire nails.

Machine and Carriage Bolts. A bolt is simply a metal rod with a head at one end and threads at the other. Bolts are manufactured in either rough or finished form. Rough bolts are most commonly used in framing work. A machine bolt and a carriage bolt's stock sizes range from lengths of $^3/_4$ to 20 in and diameters of $^3/_{16}$ to $^3/_4$ in. When longer bolts are needed and are not on hand, they may be made up in the shop or field. Rod material can be threaded or can be welded to shorter threaded bolts.

Machine Screws and Stove Bolts. Smaller bolts suitable for attaching lighter materials in heavy framing structures are machine screws and stove bolts. Machine screws come in both coarse and fine thread, while stove bolts come in the coarse thread only. Screws vary in length from $^1/_8$ in to 3 in, and stove bolts from $^1/_8$ in to 6 in. In sizes over 3 in, stove bolts are threaded to a length of 2 in. Square and hexagon nuts are available for both.

Sheet Metal Screws. Sheet metal screws are suitable where sheet metal is to be fastened together without forming seams, soldering, or riveting, or for fastening other things to sheet metal. Sizes range from $^1/_8$ to 2 in in length and from number 2 to number 14 screw gauge. The larger sizes are driven into drilled or cleanly punched holes, while the smaller sizes can be driven into pierced holes in lighter-gauge metals.

Lag Bolts. A lag bolt is basically a large screw with a square head that is turned with a wrench. The available sizes range from $^1/_4$- to 1-in thickness and from 1 to 16 inches in length. They may have a regular screw thread or a ratchet thread, which has a sharper angle. Installation requires the boring of a hole equal to the shank in diameter and depth, and another hole beyond this depth with a diameter not greater than that of the root of the thread.

A lag bolt can by used in freshly poured concrete simply by wrapping the threads with wire and placing the assembly in position. When the concrete has set, the lag bolt can be unscrewed, leaving the wire embedded in the concrete for a permanent thread.

Dowel Screws. A bolt or rod threaded like a screw at both ends is called a dowel screw, which is used where high strength is not required. Holes are needed for installation much as with a lag or coach screw. In joining two pieces together with a dowel screw, it is necessary to turn one or both pieces.

Handrail Bolts. A handrail bolt is similar to the dowel screw except that the threads accommodate nuts at each end. A stud bolt is also threaded at both ends; however, there is usually more thread at one end than at the other.

Eye Bolts. These are bolts with eyes formed by a loop at one end; the other end is threaded. The eye is convenient for fastening cables.

Turnbuckles. These are devices threaded right-handed at one end and left-handed at the other. They are used to take up slack in rods or to tighten cables.

Washers

Washers are used in conjunction with bolts in structural framing to increase the bearing area under the head and so that the full tensile strength of the bolt can be utilized. Some of the different types of washers used in modern framing construction are: cut, malleable iron, cast iron ogee, cast iron ribbed, square steel plate, and beveled cast iron. The washer checklist is as follows:

Cut Washer. The ordinary cut washer, made of pressed steel, is relatively thin and does not provide for the maximum tensile strength of the bolt. Under stress, it may bend or permit crushing of the fibers of the wood under the washer, particularly in soft woods. The diameter is usually from two to three times the bolt diameter. Cut washers are suitable in attaching lighter materials to a heavier member.

Malleable Iron Washer. The malleable iron washer is rigid and firm but has its base area reduced in bearing because its bottom plate is not solid. The area between the outside rim and the central spool is open and joined by webs or ribs.

Cast-Iron Ogee Washer. The cast-iron ogee washer has a diameter about four times that of the bolt with which it is used. The larger sizes are to be used with soft woods. The hole tapers, being larger on the top or face by $^1/_8$ in. in the smaller sizes and $^1/_4$ in. in the larger. The thickness of the washer is usually either the same as the bolt diameter or $^1/_8$ in less.

Cast-Iron Ribbed Washer. The cast-iron ribbed washer has a flat base plate but is ribbed on top to provide strength as well as to conserve material. It is larger than the ogee washer for the same bolt sizes, so it can be expected to carry higher loads.

Square Steel-Plate Washers. Square steel-plate washers may be readymade or cut out on the job. Their efficiency depends on the size of the washer and its thickness.

Beveled Cast-Iron Washer. Beveled cast-iron washers are made to hold a bolt where it projects through the timber at an angle with the surface.

Cup Washers. Cup washers are designed for the holes of countersunk bolts. *Socket washers* serve the same purpose in construction like the floors of elevators and freight cars.

Segmental Washer. A segmental washer is two identical malleable iron castings that fit together to form a simple round washer. The castings will not separate when the bolt is tightened against them. Since the washer can be installed without removing the bolt nut, its use is valuable when threads on the bolt run out or become rusted.

Fastener Location

To properly connect sheathing to framing, most nails should be located in the center of the framing member whenever possible. Centering the nail is easy to do on sill plates, sole plates, and intermediate studs. Centering nails in the uppermost top plate frequently requires oversized panel sheets.

Centering nails in framing members or blocking at adjoining panel edges is impossible. To make up for the inability to center the nails at adjoining panel edges, wider framing members and blocking are recommended at adjoining panel edges.

When nails are installed too close to the sheathing edge, the shear wall will fail prematurely during seismic loading. The greater the edge distance, the better the sheathing connection will perform. High-strength diaphragms are created when several rows of nails are placed in wide framing members with large edge distances. Although current building codes require only $^3/_8$ in, the minimum recommended edge distance for sheathing fasteners is $^1/_2$ in. This will increase the strength of the connection.

Many builders and inspectors have expressed confusion about the actual meaning of the code language that states nail heads are "not to fracture the surface of the sheathing.

The head or crown of the nail shall be flush to the surface of the sheathing." The problem here is that nails installed with their heads resting on but not into the sheathing can cause problems for roofing and other finishes. Nails should be driven so that the top of the nail is flush and not above the surface of the sheathing.

Because nails will try to pull through the thickness of wood structural panel sheathing during an earthquake, nails should be driven flush with the surface and not overdriven. Overdriven nails reduce the shear wall strength by effectively reducing the thickness of the sheathing. At panel edges, overdriven nails allow easier nail punch-through.

At intermediate studs, overdriven nails allow easier panel buckling. Nails are commonly overdriven because of excessive pressure or too long a driving pin on the nail gun. Hand-driven nails are rarely overdriven unless a soft spot in the framing member or blocking exists where the nail is installed.

Joists and Girders

Joists are the interlocking horizontal framing members that tie the mudsill at the foundation walls to the floor, which the joists also support. The most common spacing of wood joists is 16 in on center (o.c.). In certain job design situations, 24 in o.c. is acceptable if the floor sheathing is designed for the structural span load on the floor joists. The code tables list safe spans for joists under average loads.

For floor joists, this is usually figured on a basis of 50 lb/sq ft (10 lb dead load and 40 lb live load). Joists must not only be strong enough to support the loads placed upon them, but they must also be stiff enough to prevent undue twisting, bending, deflection, vibration, or uplift. See Figure 1-10.

The building codes specify that the deflection (bending downward at the center) must not exceed $\frac{1}{360}$th of the span with a normal live load. This would equal $\frac{1}{2}$ in for a 15-ft span. The dispersed loads on each joist would then be similar to Fig. 1-11.

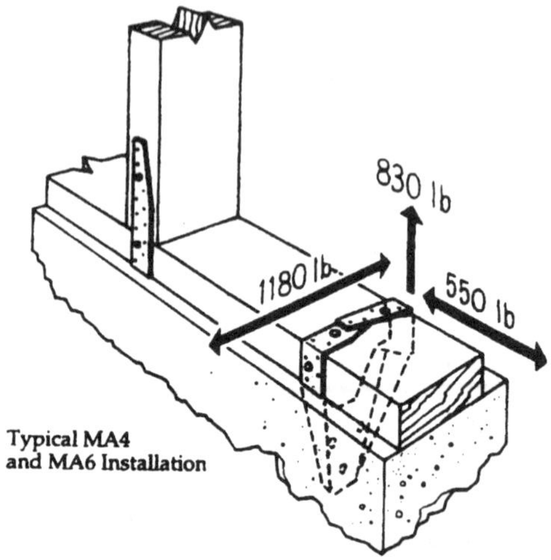

Figure 1-10 Three-way Floor Loads

Typically, because the span between loadbearing foundation walls is so great, support members are installed at the halfway and quarterway points. These loadbearing members are either girders or grade beams. Girders may be either solid wood, built-up laminated lumber, or steel beams. If the ground is relatively flat, a concrete footing tied into the perimeter footing can act as another loadbearing stemwall and replace the need for girders at that point.

Joists under parallel bearing partitions must be doubled or a beam of adequate size to support the load must be provided. Double joists that are separated to permit the installation of piping or vents must be solid blocked and spaced not more than 4 ft (1219 mm) on center. The ends of each joist, beam, or girder must have not less than 1 $^1/_2$ in (38 mm) of bearing on wood or metal and not less than 3 in on masonry except where supported on a 1 in × 4 in (25 mm × 102 mm) ribbon strip and nailed to the adjacent stud or by use of approved joist-to-masonry connectors. See Figure 1-12.

Floor systems having joists framing from opposite sides over a bearing support must be tied together by lapping the joists a minimum of 3 in (76 mm) or with a wood or metal

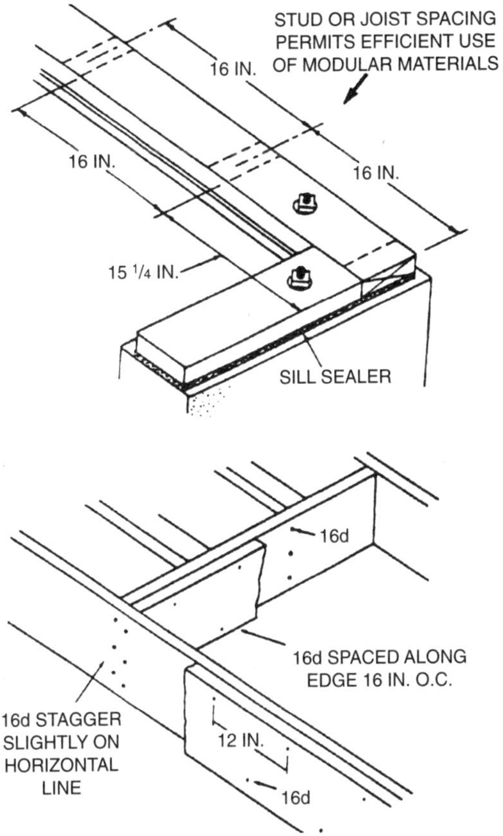

Figure 1-11 Laying Out Joists

splice connector, or by continuity of floor sheathing overlapping the ends of joists at least 3 in (76 mm) or by other approved methods. Joists framing into the side of a wood girder must be supported by approved framing anchors or on ledger strips not less than nominal 2 in × 2 in (51 mm × 51 mm). Floor joists must also be supported laterally at the ends by full-depth solid blocking not less than 2 in (51 mm) in thickness or by attachment to a header, band, or rim joist or adjoining stud, or be otherwise provided with lateral support to prevent rotation (Fig 1-13). This lateral support is

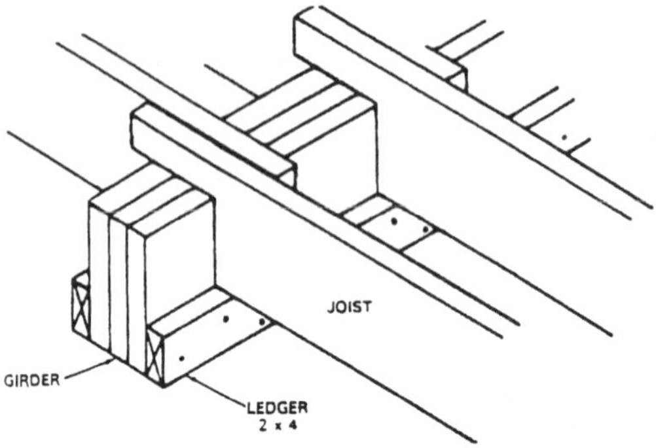

Figure 1-12 Framing Floor-Joists

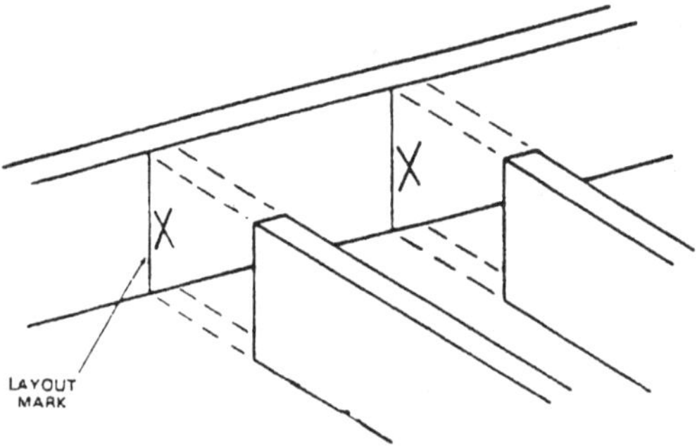

Figure 1-13 Framing Floor-Joists

not required over intermediate supports such as center girders or bearing walls.

Joists having a depth-to-thickness ratio exceeding 6 to 1 based on nominal dimensions must be supported laterally by solid blocking, diagonal bridging (wood or metal), or a continuous 1 in × 3 in (25 mm × 76 mm) strip nailed across the bottom of joists perpendicular to joists at intervals not exceeding 10 ft (3048 mm).

To determine the structural size needed of a girder, first measure the distance between girder supports. A girder must bear the weight of the floors on each side to the midpoint of the joists that rest upon it; this is called *girder load*. Determine the girder load. Then determine the total floor load per square foot carried by the joists and bearing partitions to the girder. (Be sure to include roof loads.) Now find the total load on the girder by multiplying the span by the total floor load and multiplying that by the girder load width. Select the proper size by cross-referencing the resulting size according to the code table in your area.

Built-up girders can be made of sistering three or four 2-in nominal width stock of the same dimensions with 20d nails. Nails in a built-up girder should not be spaced farther apart than 32 in along top and bottom edges, and staggered to prevent splitting. Metal bearing plates should be installed under the girder at foundation walls. Table 1-1 shows a standard code-compliant girder size and load table.

Beams and girders intersections are subject to the forces shown in Fig. 1-14. The position of the floor joists can be laid out directly on the sill plate; however, in platform framing the joist spacing is usually laid out on the joist header rather than the sill plate. The position of an intersecting framing member is laid out, and all crowns are turned upward.

Instead of measuring each individual space around the perimeter of the building, it is more accurate and efficient to make a master layout on a strip of wood (called a rod). Use it to transfer the layout to the headers or sill plate. The same rod is then used to make the joist layout on the girders and the opposite wall (Fig 1-15).

Table 1-1
Girder Sizes and Generally Accepted Loads

Girder Size	Safe Load in lbs (Spans 6 - 10 ft)				
	6 ft	7 ft	8 ft	9 ft	10 ft
6×8 solid	8,306	7,118	6,220	5,539	4,583
6×8 built-up	7,359	6,306	5,511	4,908	4,062
6×10 solid	11,357	10,804	9,980	8,887	7,997
6×10 built-up	10,068	9,576	8,844	7,878	7,086
8×8 solid	11,326	9,706	8,482	7,553	6,250
8×8 built-up	9,812	8,408	7,348	6,544	5,416
8×10 solid	15,487	14,782	13,608	12,116	10,902
8×10 built-up	13,424	12,768	11,792	10,504	9,448

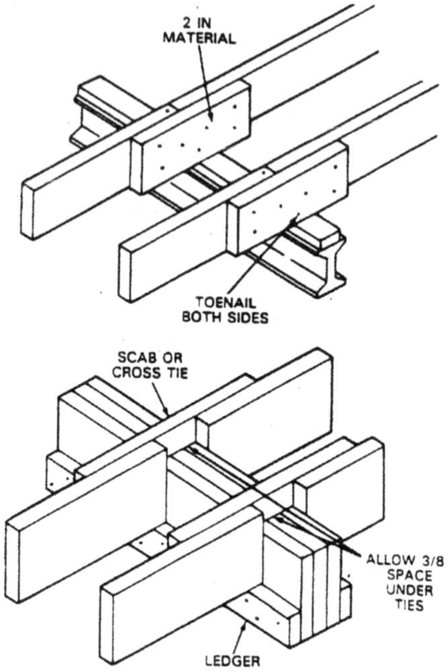

Figure 1-14 Framing Over Girders

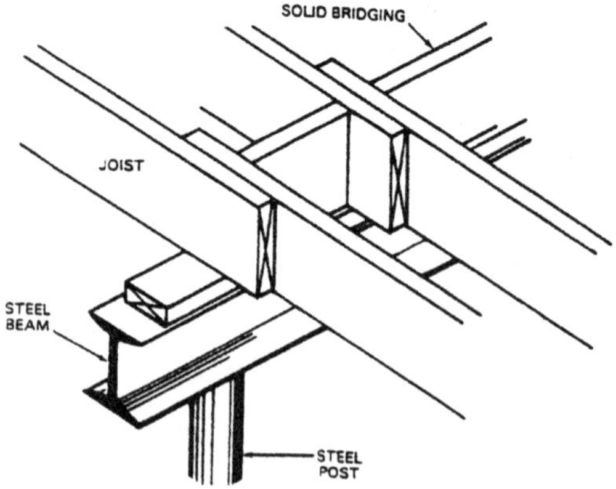

Figure 1-15 Framing Over Beams

Joists are doubled where extra loads must be supported. When a partition runs parallel to the joists, a double joist is placed underneath for additional support. Partitions which carry plumbing or heating pipes are usually spaced far enough apart to permit easy access. Joists must also be doubled around openings in the floor for stairways, chimneys, and fireplaces. These joists are called *trimmers* and support the headers that carry the tail (short) joists.

Design criteria for structural framing members such as joists and rafters are available in the local adopted building code. The standards for fastening are listed in Table 1-2.

Pintles

Pintles are used to transfer loads of columns on upper floors through intervening beams and girders to support column caps. The pintle serves as a base for the column on one floor and is used to avoid problems of shrinkage that

Table 1-2
Framing Connections and Their Generally Accepted Nail Sizes for Fastening

Framing Connection	Common Nail Size
Stud to sole plate	4-8d toe nails or 2-16d end nails
Stud to top plate	2-16d end or toe nails
Double studs	10d face nails 12 in o.c.
Trimmers	2-16d end nails
Corner studs	16d face nails 24 in apart
Sole plate to joist	16d face nails 16 in apart
Double top plate	10d face nails 16 in apart
Lap joints in top plate	2-10d face nails
Rafter to top plate	3-8d toe nails
Rafter to ridge beam	2-16d end or toe nails
Jack rafter to hip rafter	3-10d toe nails or 2-16 end nails
Floor joist to sill	3-8d toe nails
Ceiling joist to top plate	3-16d toe nails
Ceiling joist lap joint	3-10d face nails
Ceiling joist to rafter	3-10d face nails
Collar tie to rafter	3-10d face nails
Bridging to joists	2-8d face nails ea. end
Let-in braces	2-8d face nails ea. stud
Tail joists to headers	Joist hangers
Headers to trimmers	Joist hangers
Ledger to header	3-16d face nails per tail joist

would cause trouble due to uneven or unequal seasoning in the lumber used.

Floor Truss Performance Checklist

Considerable research related to human response to movements and/or vibrations in floor systems has been conducted. While this research has not resulted in revised guidelines regarding deflection limitations, it has underscored the need to provide relatively stiff floor systems to overcome human objections to floor movements. The dynamic loadings and subsequent responses encountered in floor systems is a complex subject, and the spectrum of human response to these dynamic effects is even more complex. Modern truss components (see Fig. 1-16) are engineered for the truss system to respond to these dynamics.

In addition to floor movements induced by loadings, noises in the floor system, such as squeaks that may occur because of movement between mechanical fasteners such as nails attaching the floor sheathing to the trusses, are also

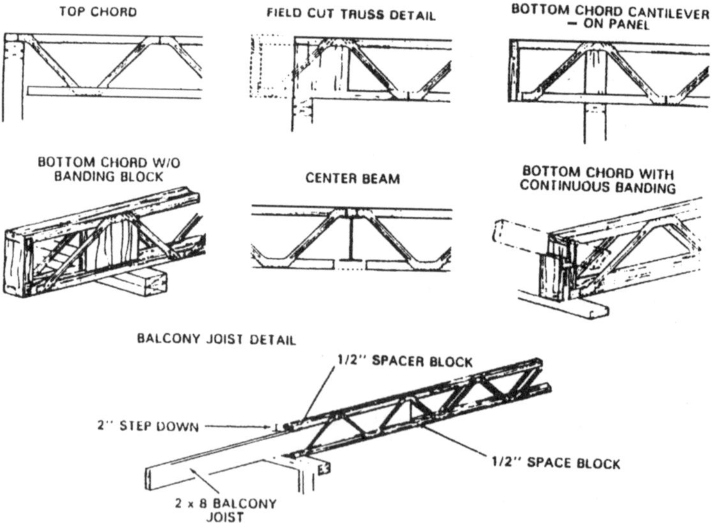

Figure 1-16 Truss Components

objectionable. To minimize these potential human objections to the performance of wood truss floor systems, the following items are suggested as a truss performance checklist to be considered by the framer.

- □ Good workmanship in the field during the stages of installation of the trusses is most critical, particularly in ensuring that all bridging and bracing are properly installed and that required bracing conditions are satisfied. This can only be ensured by adequate field inspection by the framing foreperson, architect, truss supplier, and building official.
- □ Total uniform live-load deflection as a criterion of performance may not be as significant as the movement and/or vibration induced in the floor by the dynamic loadings of a person walking across the floor. Applying conventional live-load deflection criteria as required by the code authorities may not eliminate this concern.
- □ A floor system which provides composite action between the sheathing and supporting trusses such as with a nailed-glued system acting in conjunction with an effective bridging system helps minimize these objectionable movements. In addition to minimizing floor movements, the nailed-glued system also results in fewer noise objections.
- □ Thicker sheathing for a given span condition provides greater stiffness between trusses with better dampening characteristics. A double-layer system or an equivalent single-layer material helps to reduce floor movements.
- □ Bottom chords of trusses should be laterally restrained to eliminate torsional vibrations and rotation.
- □ An unloaded or lightly loaded floor system generally does not perform as well as a more heavily loaded system because of the dampening effect that the loads have on floor vibrations.
- □ Structural discontinuities between separate living occupancies in multioccupancy structures should be provided.

- □ Carpeting or other resilient floor coverings help reduce objectionable vibrations and noise. However, they also can aggravate the movement of unattached less stable objects on the floor.
- □ In general, the human sensitivity to floor performance increases as the escalating cost and sophistication of the use of the building increases.
- □ As a general rule, once people become sensitized to the poor performance of a floor system, it is very difficult to desensitize them through remedial modifications of the system. Thus the initial human response to a floor system becomes of significant importance.

The preceding checklist relates primarily to field experiences rather than specific engineering issues and, as such, is not addressed by the building code authorities. However, how a floor performs with respect to human perception of that floor's response to dynamic loads is a critical design consideration of the truss system performance.

Floor Framing Checklist

The load upon the floor framing is determined by the accumulative totals of the entire load path, from top to bottom, throughout the structural framing system. In sequence, they are:

- □ Live load on roof = local code requirements for wind and snow (usually 30 lb/sq ft)
- □ Dead load of roof = roofing (wood shingles = 10 lb/sq ft)
- □ Live load on attic floor = local code requirements (usually 20 lb/sq ft when used for storage only)
- □ Dead load of attic floor (not floored) = 10 lb/sq ft
- □ Dead load of attic floor (floored) = 20 lb/sq ft (storage only)
- □ Dead load of partitions = 20 lb/sq ft of floor area
- □ Live load on second floor = local code requirements (usually 40 lb/sq ft)

- □ Dead load on second floor = 20 lb/sq ft
- □ Dead load of partitions v 20 lb/sq ft of floor area
- □ Live load on first floor = local code requirements (usually 40 lb/sq ft)
- □ Dead load of first floor (ceiling not plastered) = 10 lb/sq ft
- □ Dead load of first floor (ceiling plastered) = 20 lb/sq ft

When a wood column or post supports a flooring girder, the post's greatest dimension must equal the width of the girder it supports. At any loadbearing intersection of the subfloor framing where wood supports wood, a metal connector plate should be installed to provide an even bearing surface. A metal cap on post ends will prevent the end grain of the post from cruching and distorting the horizontal grain of the wood girder.

Floor framing consists of either a truss system like the one shown in Fig. 1-17 or joists that are supported by the girders on top of the grade beam that is set flush with the top of the foundation wall. Installing the floor joists to the grade beam at various levels is accomplished with special hangers. Joist and beam hangers are used when the bottom of the girder must be flush with the bottoms of the joists. Allow for shrinkage when framing the floor joists. For average work with a 2 × 10 in joist, an allowance of $^3/_8$ in above the top flange of the grade beam is usually sufficient.

When installing the foundation sill plate, remove the washers and nuts from the anchor bolts. Lay the sill along the foundation wall and draw 90° lines across the sill plate from the bolts. Remember to account for the edge of the sill that will be set back from the outside of the foundation a distance equal to the thickness of the sheathing (Fig. 1-18). Measure the distance from the center of the bolt to the outside of the foundation and subtract the thickness of the sheathing.

Use this measurement to locate the bolt holes to be drilled in the sill plate. Because it's not a perfect world, you will probably need to make separate measurements for each anchor bolt. And since foundation walls are seldom perfectly

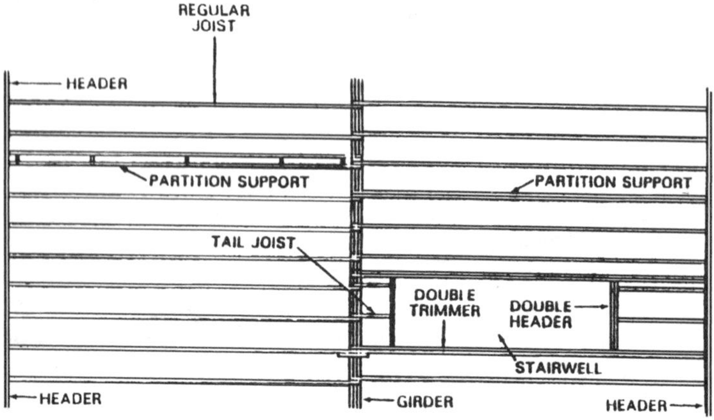

Figure 1-17 Floor Framing

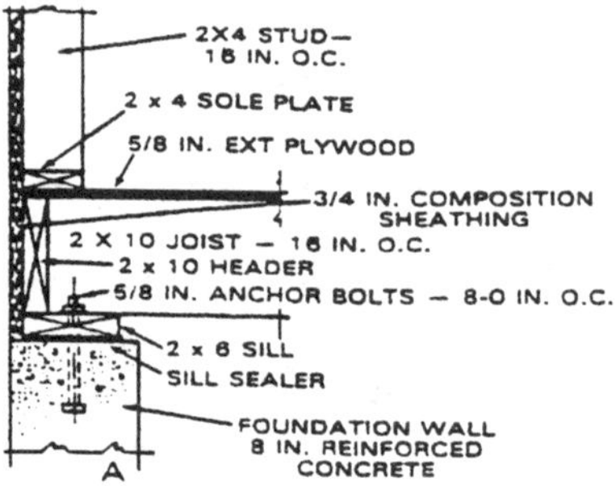

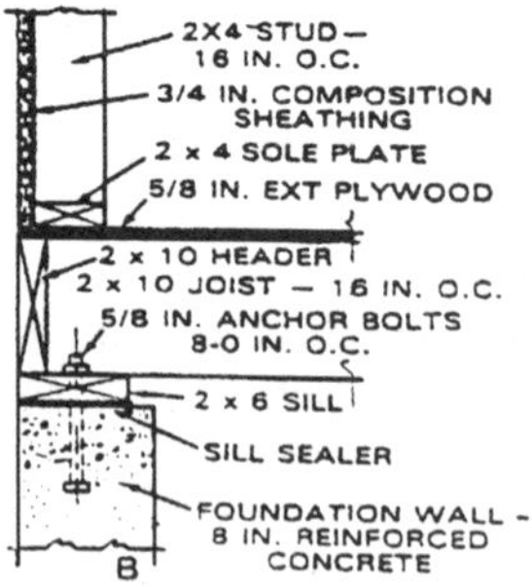

Figure 1-18 Framing at Sill

straight, many framers prefer to snap a chalk line along the top where the outside edge of the sill should be located. This will ensure an accurate floor frame that is basic to all additional construction.

Variations between the outside surface of the sheathing and foundation wall can be shimmed when the siding is installed.

Structural Framing Diaphragms

In structural framing, a "designated seismic system" is the seismic resisting system and those engineered loadbearing framing components that require special performance characteristics. To be effective, each part of the designated seismic system must be adequate and properly connected to the other parts in the framing system. For example, the roof diaphragm must be strong enough to safely resist the seismic loads.

Then the connections from the roof diaphragm to the shear walls below must be strong enough to give the force to the shear wall. Shear walls at the base of the building's frame must be connected to the building foundation's footing, and the footing must be adequate. To have an effective horizontal force-resisting system, there must also be a continuous load path throughout the framing structure as shown in Fig. 1-19.

Diaphragms are horizontal, or nearly horizontal, portions of the seismic resisting system, which is designed to transmit seismic forces to the vertical elements of the seismic resisting system. The vertical elements are the walls that span between the horizontal elements, which are the shear walls. Wood structural diaphragms have ties or struts to distribute the wall anchorage forces into the diaphragm. All framing diaphragm connections must be positive, mechanical-type connections.

The seismic load path can be thought of as a chain that is only as strong as its weakest link. The roof and floor diaphragms and shear walls are two links in the chain. The connectors are just as important as the diaphragms and

Complete Load Path

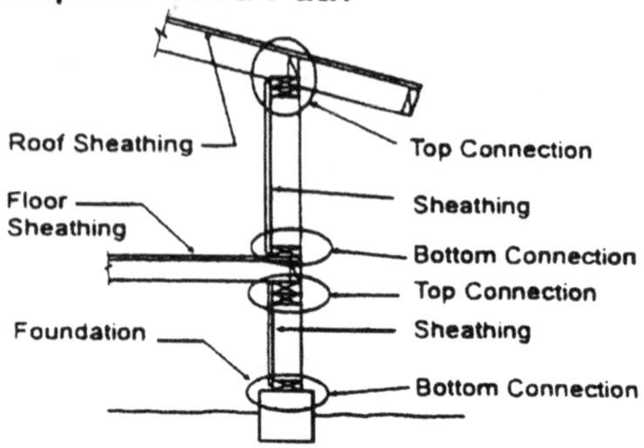

Figure 1-19 Load Path

shear walls themselves. The seismic loads imparted on a building must successfully pass through all these elements in order to reach the ground and effectively resist an earthquake's damaging forces.

In other words, the load path must be continuous and complete. There can be no weak links in the load path chain. The entire shear load path chain through two floors of framing is shown in Figs. 1-20 and 1-21, numerically sequenced from roof to ground.

When a part of the load path is missing, damage will occur during an earthquake. If the sill plate is not connected by the use of anchor bolts to the foundation wall, the ground will move and leave the building behind. Buildings without shear walls can collapse. When there is not enough room for the shear wall, an engineer or architect must design a special frame to resist the horizontal forces. These frames are called *moment frames*.

The diaphragms in buildings having torsional irregularity, where the lateral stiffness ratio of the structural members is greater than 4:1, or in buildings with one line of resistance in either orthogonal direction are sheathed with diagonal boards or wood structural panels. The length of the diaphragm normal to the soft side cannot exceed 25 ft (7.6 m) nor can the diaphragm length-to-width ratio exceed 1 for one-story frame or 0.67 for buildings over one story in framing height.

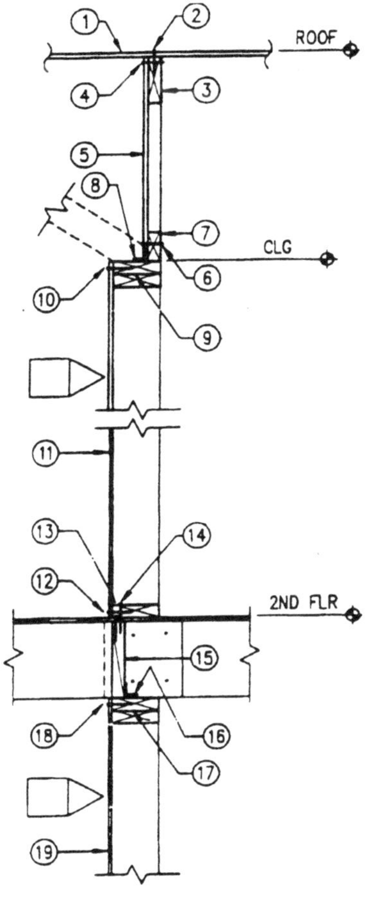

Figure 1-20 Second Floor Sequential Load Path

Where engineering calculations can demonstrate that the diaphragm deflections will be tolerated, the length limit of 25 ft (7.6 m) does not apply, and the length-to-width strength ratio of 1.5 is permitted for framing diaphragms sheathed with single diagonal boards and a ratio of 2 where sheathed with double diagonal boards or wood structural panels.

The deflection in the plane of the diaphragm, as determined by engineering analysis, cannot exceed the permissible deflection of the attached framing elements. In terms of framing, permissible deflection is that deflection that will

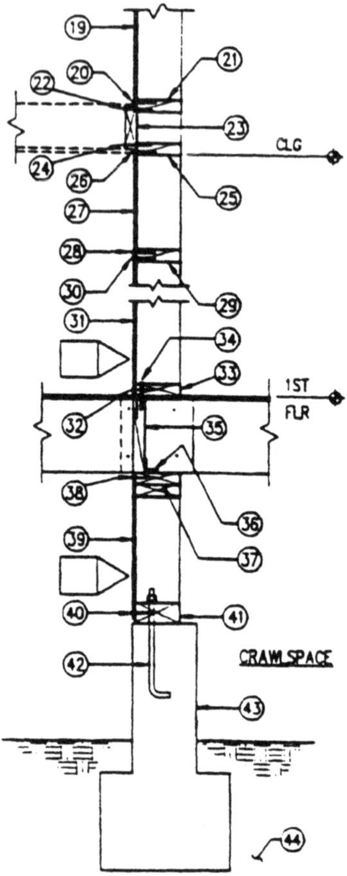

Figure 1-21 First Floor Sequential Load Path

permit the attached framing element to maintain its structural integrity under the individual seismic loading forces and continue to support the normal prescribed design loads.

The design shear capacity is permitted to be doubled when identical materials are applied to both sides of the wall. Where the shear capacities of the materials are not equal, the allowable shear must be considered to be equal to either the shear for the side with the higher capacity or twice the shear for the side with the lower capacity.

Floor and roof diaphragms are be designed to resist the following seismic forces listed in the building code tables: A minimum force equal to 50 percent the effective peak velocity-related acceleration A_V, times the weight of the diaphragm and other elements of the building attached thereto plus the portion of the seismic shear force at that level V_X required to be transferred to the components of the vertical seismic resisting system because of offsets or changes in stiffness of the vertical components above and below the diaphragm. Structural diaphragms provide for both the shear and bending stresses resulting from these forces.

Gypsum wallboard, gypsum sheathing, and gypsum veneer base may be used on wood studs for vertical diaphragms when such construction is designed to resist seismic forces if applied in accordance with local adopted code. However, shear resisting values, panel size, and arrangement provisions cannot exceed those set forth in the adopted code tables.

The shear values tabulated cannot be cumulative with the shear value of other materials applied to the same wall. Cumulative allowable shear values for walls sheathed with more than one type of material must be supported by engineering calculations or tests. The shear values may be doubled when identical materials applied as specified in the adopted code are applied to both sides of the wall.

Framing for vertical diaphragms must comply with specifications for bearing walls. Studs must be spaced no farther apart than 16 in (406 mm) center to center. Marginal studs and plates must be anchored to resist all design forces, and the maximum allowable height-to-length ratio for the construction is $1^1/_2$:1. End joints of adjacent courses of gypsum board sheets cannot occur over the same stud.

Where required by adopted code, blocking having the same cross-sectional dimensions as the studs must be provided at all joints that are perpendicular to the studs. The spacing of nails cannot be less than $^3/_8$ in (9.5 mm) from edges and ends of gypsum boards or sides of studs, blocking, and top and bottom plates. Gypsum sheathing 4 ft (1219 mm) wide may be applied parallel or perpendicular to studs.

Blocked Diaphragms

A blocked diaphragm is a structural shear diaphragm in which all sheathing edges not occurring on a framing member are supported on and connected to blocking. All wood columns supporting such blocking must be provided with full end bearing, with adequate support to maintain stability. Where post-and-beam or girder construction is used, positive connections are provided to resist uplift and lateral displacement.

All framing members used in blocked diaphragm shear panel construction must be at least 2 in nominal thickness. Boundary members, chords, and collector members must be constructed to transfer the axial forces. Boundary members must be connected at all corners. Openings in shear panels must be designed and constructed to transfer all shear loads into the blocked diaphragm. Positive connections and anchorages, capable of resisting the design forces, must be provided between the shear panels and the attached components. These design forces move in three directions resisting compression, tension, and shear forces. Examples of the framing connections that will resist in three directions as required are shown in Fig. 1-22.

Shear Walls

As stated in the previous section on diaphragms, shear walls are vertical elements of the frame's horizontal forces resisting system. They are typically wood frame stud walls covered with a structural sheathing material like plywood. When the sheathing is properly fastened to the stud wall framing, the shear wall can resist forces directed along the length of the wall. When shear walls are designed and constructed properly, they will have the strength and stiffness to resist the horizontal forces. See Fig. 1-23.

The recommended panel grade for shear wall construction is Structural 1. Structural 1 is the premium grade of both CDX and OSB panels. Structural 1 rated sheathing has increased cross-panel strength and shear properties, and all plies use special improved grades. The allowable panel

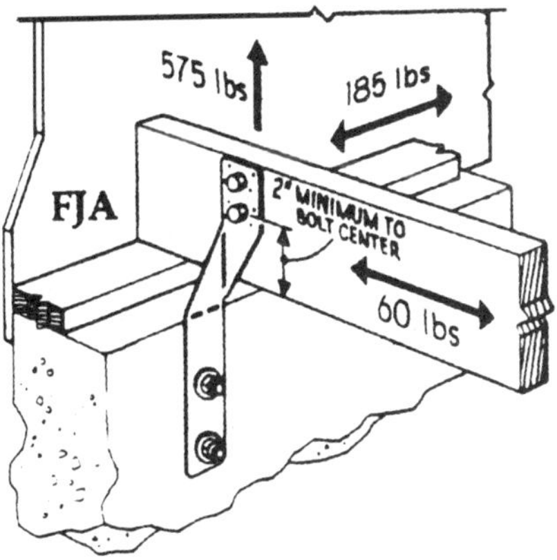

Figure 1-22 Three Directions of Force at Joist

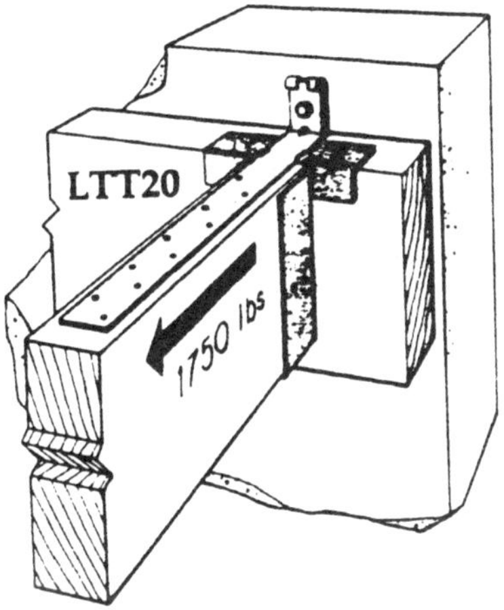

Figure 1-23 Horizontal Force

rigidity of Structural 1 is almost twice that of any other structurally rated sheathing. Because of their properties, Structural 1 shear walls are 10 percent stronger than other similar structural use sheathing.

Shear walls should be located on each level of the structure including the crawl space. To form an effective box structure, equal length shear walls should be placed symmetrically on all four exterior walls of the building. Shear walls should be added to the building's interior when the exterior walls cannot provide sufficient strength and stiffness or when the allowable span-width ratio for the floor or roof diaphragm is exceeded. For subfloors with conventional diagonal sheathing, the span-width ratio is 3:1. This means that a 25-ft-wide building with this subfloor will not require interior shear walls until its length exceeds 75 ft unless the strength or stiffness of the exterior shear walls is inadequate.

Shear walls are most efficient when they align vertically and are supported on foundation walls or footings. When shear walls do not align, other parts of the structural frame will need additional strengthening. Consider the common case of an interior wall supported by a subfloor over a crawl space where there is no continuous footing beneath the wall.

For this foundation cripple wall to be used as a shear wall, the subfloor and its connections will have to be strengthened through the wall. See Fig. 1-24 for an example of a typical sheathing load path. For new construction, thicker plywood or extra nailing and connections can be added. For retrofit work, existing floor construction is not easily changed. That's the reason why most retrofit work uses walls with continuous footings underneath them as shear walls.

Another type of alignment problem occurs when the ends of shear walls do not align from story to story. This condition creates the need for extra framing members and connections in the walls for holdown devices. Holdown devices must transfer the uplift from the shear walls to framing members that can resist it. When full height studs are not available, special connections must then be added. These connections must assemble enough of the structure's framing to resist what is termed *uplift force*.

Shear walls resist two types of forces: shear forces and uplift forces. Connections to the framing above transfer horizontal forces to the shear wall. This transfer creates shear forces throughout the height of the wall between the top and bottom shear wall connections. The strength of the lumber, sheathing, and fasteners must resist these shear forces or

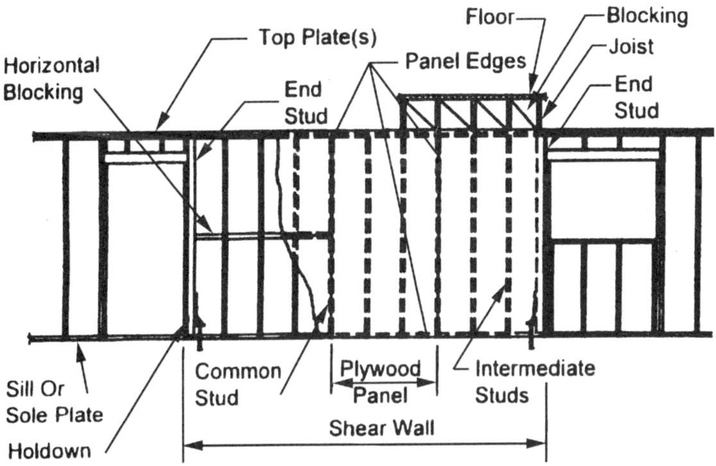

Figure 1-24 Shearwall Framing

the wall will tear or "shear" apart. Uplift forces exist on shear walls because the horizontal forces are applied to the top of the wall.

These uplift forces try to lift up one end of the wall and push the other end down. In some cases, the uplift force is large enough to tip the wall over. Uplift forces are greater on tall short walls and less on low long walls. Bearing walls have less uplift than nonbearing walls because gravity loads on shear walls help them resist uplift. Shear walls need holdown devices at each end when the gravity loads cannot resist all of the uplift. The holdown device then provides the necessary uplift resistance.

Shear walls must also provide the necessary lateral strength to resist horizontal earthquake forces. When shear walls are strong enough, they will transfer these horizontal forces to the next element in the load path below them. These other components in the load path may be other shear walls, floors, foundation walls, slabs, or footings. Shear walls must also provide the necessary lateral stiffness to prevent the roof or floor above from excessive side-sway. When shear walls are stiff enough, they will prevent floor and roof framing members from moving off their supports. Also, framing that is sufficiently stiff will usually suffer less nonstructural damage.

As of this writing, the building codes still allow the use of gypsum wallboard, cement plaster, fiberboard, wood particleboard, and oriented strand board (OSB) for shear wall sheathing. However, the codes are in the process of restricting their use in some applications of shear wall sheathing. All three of the major model building codes plan to replace the applications of these products as shear wall sheathing in the near future with mandatory plywood. Fasteners for shear wall construction may be staples, screws, or nails; used in conjunction with metal splice straps going either perpendicular or parallel to the wall (see Figs. 1-25 and 1-26). Denser lumber species provide stronger fastener strengths. Values for shear wall strengths assume a dense lumber species like Douglas Fir-larch or Southern Pine. Thicker framing members also increase wood structural panel sheathing strengths.

When nailing shear wall sheathing, framers should operate their nail guns within the manufacturer's recommended pressure range and install flush nailing attachments as necessary. Inspectors should reject all nails driven below the surface of the sheathing as this perforates and defeats the shear wall. When the spacing and framing member thickness allow, nails should be added to replace any overdriven nails.

Shear Wall Framing

Most 8-ft walls today are framed at a height of 8 ft $^3/_4$ in to allow for the gypsum wallboard on the ceiling. As a result, the common 8-ft sheet of shear wall sheathing does not span the full height of the wall. Frequently sheathing panels are installed vertically with one end flush to the bottom edge of the sill or sole plate and the other end centered on the uppermost top plate.

This situation cannot easily be fixed by simply adding 3-in top plates. Thicker plates require longer framing nails that are not readily available and are more difficult to install.

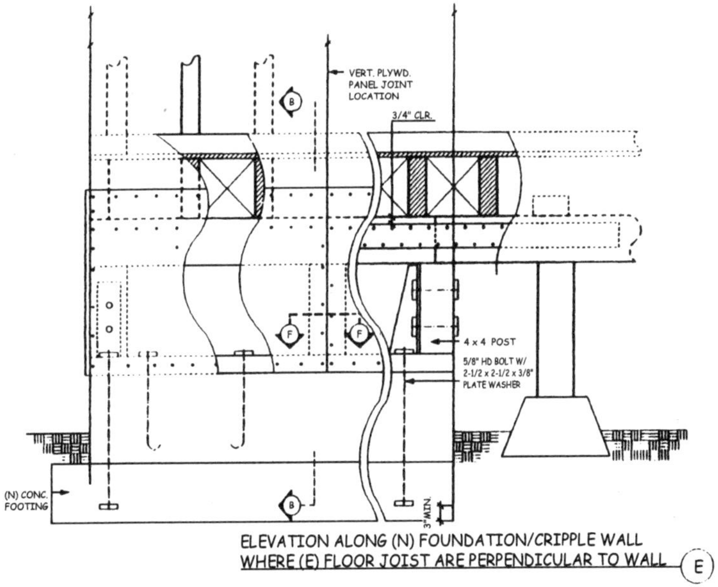

Figure 1-25 Perpendicular Splice Straps

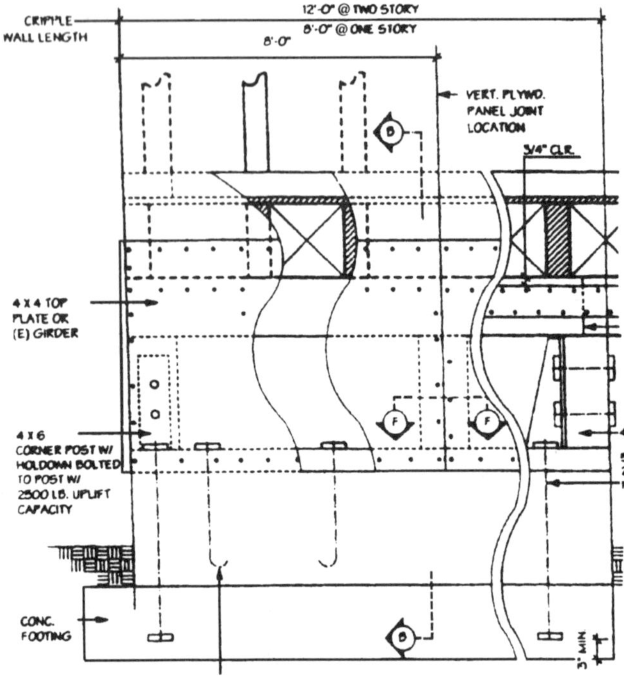

Figure 1-26 Parallel Splice Straps

The best solution is oversized sheets. Both OSB and plywood come in 9- and 10-ft lengths, and when these longer lengths are used, shear transfer connections can be eliminated if nailing is provided to the floor- and roof-framing members.

In shear wall framing, it is critical to check for proper nailing in the connection of the foundation to floor in the different types of modern foundation stemwalls. See Figs. 1-27 and 1-28. All fasteners in wood structural framing must be installed without splitting the lumber. There are no exceptions. Existing lumber will often be dry and stiff, so if needed to prevent splitting, predrilled holes at 75 percent of the nail diameter are required. For the shear wall to reach its intended strength, all sheathing edges of individual panels must be fastened with the local code edge distance nailing requirements.

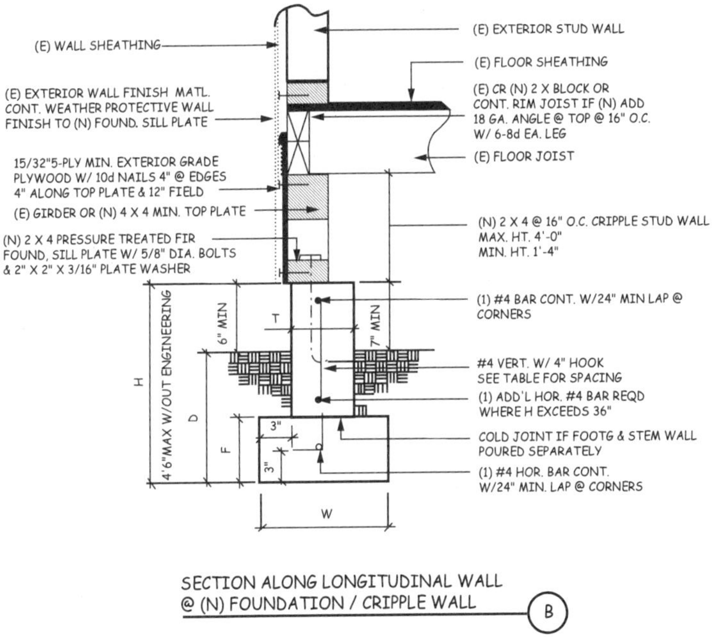

Figure 1-27 Stemwall to Frame Connection

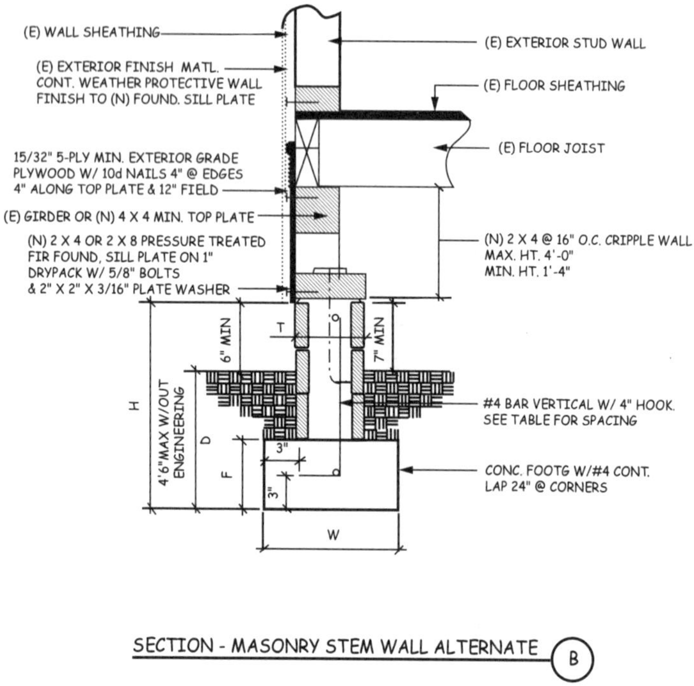

Figure 1-28 Masonry Stemwall

Intermediate studs must be fastened with the field nailing requirements so that panel buckling does not occur. Remember that the end of a shear wall is where the sheathing stops at either wall corner or the framing member with the holdown device. All shear wall sheathing must be fastened with the closer edge distance spacing at every framing member that connects to a holdown device.

Wood Structural Shear Panels

A wood structural shear panel is a wood floor, roof, or wall component sheathed to act as a shear wall or diaphragm. The design and shear capacity of wood structural shear panels are calculated by using nail strengths in the NFiPA NDS and plywood design properties as given in the

American Plywood Association (APA) Plywood Design Specifications.

Wood structural shear panels are constructed of wood structural panels, manufactured with exterior glue, not less than 4 × 8 ft (1.22 × 2.44 m), except at boundaries and changes in framing. The minimum recommended thickness for wood structural panels is $^{15}/_{32}$ in. For this thickness, both four-ply and five-ply layers are commonly available. Five-ply panels provide five plies and five layers while four-ply panels provide four plies and three layers.

Wood structural panels are designed to resist shear only, and chords, collector members, and boundary members are designed to transfer the axial forces. Boundary members are connected at all corners. Wood structural panels less than 12 in (305 mm) wide must be blocked. Wood structural panels used for shear panels that are a part of the seismic-load-resisting system must be applied directly to the framing members. The only exception are wood structural panels nailed over solid lumber decking or laminated decks.

Single diagonally sheathed lumber shear panels must be constructed of minimum 1-in-thick nominal sheathing boards laid at an angle of approximately 45° (0.785 rad) to the supports. The shear capacity for single diagonally sheathed lumber shear panels of Southern Pine or Douglas Fir-larch cannot exceed 200 lb/lin ft (2.92 kN/m) of width.

The shear capacities are adjusted by reduction factors of 0.82 for Group III framing species and 0.65 for Group IV framing species, as contained in the NFiPA NDS. End joints in adjacent boards are separated by at least one stud or joist space, and there must be at least two boards between joints on the same support.

Wood shear panels made up of 2-in nominal diagonal lumber sheathing fastened with 16d nails are designed with the same shear capacities as shear panels using 1-in boards fastened with 8d nails, provided there are no splices in adjacent boards on the same support and the supports are not less than 4 in nominal depth or 3 in nominal thickness.

Double diagonally sheathed lumber shear panels must be constructed of two layers of diagonal sheathing boards laid at 90° (1.57 rad) to each other on the same face of the supporting members. Each chord must be considered as a beam loaded with uniform load per foot equal to 50 percent of the unit shear due to diaphragm action. The load must be assumed as acting normal to the chord in the plane of the diaphragm in either direction.

The span of the chord (or any portion thereof) must be the distance between framing members of the diaphragm such as the joists, studs, and blocking that serve to transfer the assumed load to the sheathing as shown in Figure 1-29. The shear capacity of double diagonally sheathed diaphragms of Southern Pine or Douglas Fir-larch must not exceed 600 lb/lin ft (8.76 kN/m) of width. The shear capacity must be adjusted by reduction factors of 0.82 for Group III framing species and 0.65 for Group IV framing species.

Plywood Sheathing

Cheaper and lighter than most solid wood, plywood is strong and flexible and thus ideal for sheathing and shear wall applications. Plywood is made of very thin layers of wood (called *plies* or *veneers*) that are aligned and glued together. The more plies and layers that are used in plywood, the more overlapping occurs of defects and inner ply seams. For this particular reason, some prescriptive standards recommend the use of five-ply construction for plywood. Other code specifications are as follows:

- 3-ply panel construction is used on $^3/_8$ in Structural 1, $^3/_8$ in C-D Exposure 1, $^1/_2$ in C-D Exposure 1
- 4-ply panel construction is used on ½ in Structural (Southern Pine) and ⅝ C-D Exposure 1
- 5-ply panel construction is used on ½ Structural 1 (Douglas Fir), ⅝ in Structural 1

The panel thicknesses shown above are nominal. Use $^3/_8$ for $^{11}/_{32}$, $^1/_2$ for $^{15}/_{32}$, and $^{19}/_{32}$ for $^5/_8$-in thickness. Structural 1 plywood in $^1/_2$-in thickness is normally available in 4-ply (Southern Pine) panel construction because it is less expensive than the

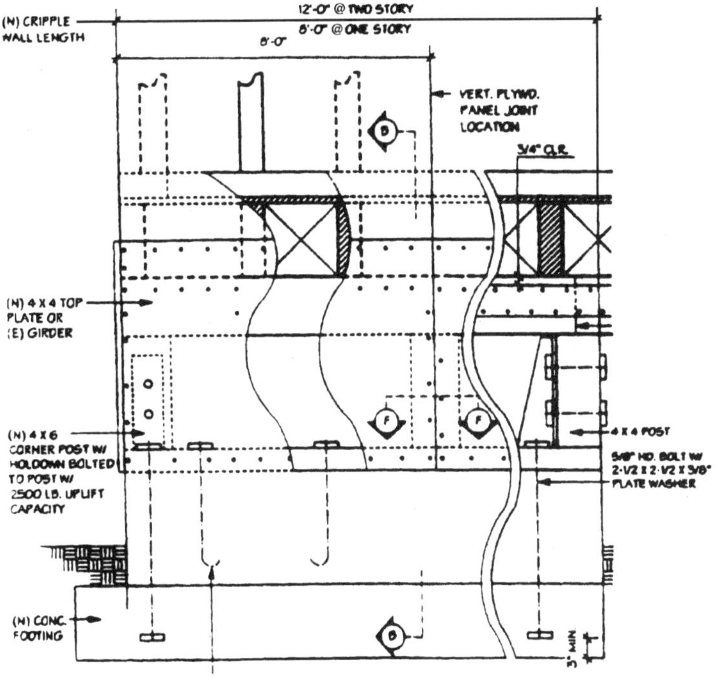

Figure 1-29 Sheathing Load Transfer

5-ply (Douglas Fir). C-D Exposure 1 allows 3-ply in $^5/_8$ in thickness, but manufacturers typically construct the panel with 4 plies to obtain the required thickness.

Generally, the middle layer of plywood, or core, is veneer, lumber, or manufactured wood. The outer plies (called *face* and *back*) are usually better appearance-quality wood than the interior ones. Face is better than back. Veneer-core plywood, the strongest type, is made up of an odd number of plies; three, five, seven, and nine plies are common compositions. Veneer-core plywood is generally available in thicknesses up to $1^1/_8$ in.

The plies are layered, with the grain of each running at right angles to those of neighboring plies. This arrangement of plies locks the grain, preventing the panel from shrinking widthwise. It also gives strength in all directions and minimizes the distortion caused by warping and shrinkage. The plies are bonded with interior or exterior glue.

Although the grains of one ply are at right angles to those of another, the panel itself does have an overall grain running lengthwise. It is this grain direction that provides the panel with the greatest strength. When installing a plywood panel, run its grain horizontal to the joists or studs.

Lumber-core plywood is made in the same way as veneer-core, except for the core, which consists of strips of solid lumber. It comes in thicknesses up to 1 in. The edges of lumber-core plywood hold nails or screws more securely than veneer-core. Plywood is further divided into two basic kinds: construction-grade and hardwood.

Construction-grade plywood, which is primarily softwood, is available in the standard 4 × 8 ft modulus of construction panel size; its common thicknesses are stocked by the local distributors at $^1/_4$, $^1/_2$, and $^3/_4$ in. Hardwood plywood, surfaced one side, is used mainly for decorative purposes and comes in panels up to 4 × 10 ft; thicknesses range from $^1/_8$ to $^3/_4$ in.

Construction-grade plywood is rated by the way it performs (span rating), regardless of thickness. Span rating refers to the maximum space (in inches) needed between supports to effectively support a panel. Most panels have one rating; sheathing has two (roof and floor). Construction-grade exterior plywood (CDX) is often used in place of wide lumber. Its plies, including its veneer core, may be softwood or hardwood or both, and it is bonded with phenol formaldehyde, an exterior adhesive.

CDX has "C" grade veneer on its face and "D" grade on its back. The "X" refers to exterior-grade use. This typical construction panel is used for subfloors and roofs. Roofs and floors generally have several panels resisting the seismic forces and therefore require lower shear strength. In contrast, shear walls generally use few panels to resist these same forces and the demand on individual panel strength and stiffness is greater than on floors or roofs.

Plywood may be manufactured from over 70 species of trees; these are divided by groups from the strongest and stiffest (Group 1) to the weakest and most flexible (Group 5). The ratings on the panel's back or edge give the group number of the face and back plies. When the face and back plies are different, the higher number is stamped.

Construction-grade plywood is manufactured in exterior and interior types. Never use interior-grade plywood for exterior applications, as weather will cause the plies to separate. Construction-grade plywood is further categorized into engineered grades and appearance grades.

The former grades are used where strength is needed in sheathing and shear wall applications and the latter in cabinetry. For construction that may be exposed to fungi and wood-destroying insects, use preservative-treated plywood and use the same precautions as for pressure-treated wood.

Table 1-3 shows the different classifications of construction plywood grades.

Particleboard Shear Panels

The design shear capacity of particleboard panels must be in accordance with local code specifications for diaphragms and shear walls. Shear panels must be constructed with particleboard sheets not less than 4 × 8 ft (1.22 × 2.44 m), except at boundaries and changes in framing.

Table 1-3
Construction Plywood Grades

Grade	Description
N	Smooth surface. Select; all heartwood or all sapwood. No open defects. No more than 6 repairs per panel. Not readily available.
A	Smooth; finishes well. No more than 18 repairs per panel. Synthetic repairs (filler or putty) permitted.
B	Solid surface. Shims, repair plugs, and tight knots permitted. Synthetic repairs permitted.
CP	Plugged. Improved C with minor splits, knotholes, and borer holes. Synthetic repairs and some broken grain permitted.
C	Small knots, knotholes, and splits within limits; synthetic or wood repairs, small defects and discoloration.
D	Larger knots and knotholes within specified limits. Larger splits permitted.

Particleboard panels must be designed to resist shear only, and chords, collector members, and boundary members must be connected at all corners. Particleboard panels less than 12 in (305 mm) wide must be blocked.

Shear Panels Sheathed with Other Materials

Wood stud wall sheathed with lath and plaster, gypsum sheathing board, gypsum wallboard, or fiberboard sheathing constructed in accordance with local adopted code specifications is, as of this writing, permitted to resist earthquake forces in wood framed buildings (Table 1-4).

The fastener location is critical to achieving design strength. Nails must be spaced at least $^3/_8$ in (9.5 mm) from the edges and ends of boards and panels, and the maximum height-to-width ratio is 1.5. The shear capacities used in construction must not be cumulative with the shear capacities of other materials applied to the same wall.

Table 1-4
Allowable Spans and Loads for Wood Structural Panel Sheating and Single Floor Grades Continuous Over Two or More Spans With Long Dimension Perpendicular to Supports[1,6]

Sheating Grade		Roof				Floor
Span Rating	Panel Thickness (in)	Maximum Span (in)		Load (psf)		
Roof/Floor Sapn		With Edge Support[2]	Without Edge Support	Total Load	Live Load	Maximum Span (In)
12/0	5/6	12	12	40	30	0
16/0	5/16, 3/8	16	16	40	30	0
20/0	5/16, 3/8	20	20	40	30	0
24/0	3/8, 7/16, 1/2	24	20[3]	40	30	0
24/16	7/16, 1/2	24	24	50	40	16[5]
32/16	15/32, 1/2, 5/8	32	28	40	30	16[4,5]
40/20	19/32, 5/8, 3/4, 7/8	40	32	40	30	20
48/24	23/32, 3/4, 7/8	48	36	45	35	24
54/32	7/8, 1	54	40	45	35	32
60/48	7/8, 1, 1-1/8	54	40	45	35	32
Single Floor Grade		**Roof**				**Floor**
Span Rating	Panel Thickness (in)	Maximum Span (in)		Load (psf)		
Roof/Floor Sapn		With Edge Support[2]	Without Edge Support	Total Load	Live Load	Maximum Span (In)
16o.c.	19/32, 5/8	24	24	50	40	16[5]
20 o.c.	19/32, 5/8, 3/4	32	32	40	30	20[4,5]
24 o. c.	23/32, 3/4	48	36	35	25	24
32 o.c.	7/8, 1	48	40	50	40	32
48 o.c.	1-3/32, 1-1/8	60	48	50	40	48

1 in = 25.4 mm
1 pst = 47 8803 Pa

Notes:

1. The allowable loads were determined using a dead load of 10 psf. If the dead load exceeds 10 psf then the live load shall be reduced accordingly.
2. Tongue-and-groove edges, panel edge clips (one midway between each support, except two equally spaced between supports 48 inches on center), lumber blocking, or other. Only lumber blocking shall satisfy blocked diagram requirements.
3. For 1/2-inch panels maximum span shall be twenty-four inches.
4. Maximum framing space shall be 24 inches on center for floors where 1-1/2 inches of cellular or lightweight concrete is applied over the panels.
5. Maximum frame spacing shall be 24 inches on center where 3/4-inch wood strip flooring is installed at right angles to joist.
6. Shall apply only to panels 24 inches or wider.

Notes

Chapter

2

Structural Wood Framing

Wood Light Frame Code Compliance

The national building codes allow buildings of every use group to be constructed with wood platform framing (which are typically classified as Type 5 construction), but there are severe restrictions on height and floor area. For some use groups, higher fire resistance ratings are required for some components of the building. These are usually achieved by using gypsum board or plaster finishes to add fire resistance to the components.

Additionally, zoning ordinances in most cities identify certain densely built areas in which wood platform frame construction is simply not permitted. And in some buildings, higher fire insurance premiums may discourage the owner from considering platform framing even if it is permitted by code.

Several examples serve to illustrate the limits within which wood platform framing can be used for different sorts of buildings. A nursing home would be classified under the BOCA Code as Use Group 1-2, Institutional, incapacitated; Type 5B wood light frame construction would not be permitted, and

Type 5A could be used for a single-story building up to 7650 sq ft (710 m^2) in floor area.

A meeting hall for a summer camp, Use Group A-3, can be only one story or 20 ft (6.1 m) tall, and 8925 sq ft (830 m^2) in area if protected (Type 5A) or 4200 sq ft (390 m^2) if unprotected (Type 5B). Retail stores (Use Group M) can be one or two stories tall with as much as 10,200 sq ft (950 m^2) of area per floor.

Hotels, apartments, and single-family dwellings may be as tall as three stories, the maximum permitted for wood platform frame construction in any use group. Buildings larger than these limits are permitted under two additional provisions of a typical building code that relate to automatic sprinklers and fire walls. The BOCA building code allows floor areas in a fully sprinklered building to be increased beyond the limits set forth in standard code tables by 200 percent for one- and two-story buildings and by 100 percent for taller buildings.

With certain exceptions for hazardous and institutional buildings, the allowable height may also be increased by one story, thus allowing a sprinklered platform frame building to rise as high as four stories.

The area of a building may also be increased above the code-defined limits by subdividing the building with *fire walls*. Under the BOCA Code, each portion of a building that is separated from the remainder of the building by fire walls may have a floor area only as large as that permitted by code tables. Every building code has a similar provision. The required fire resistance of the fire wall is given in a separate table in the code.

A fire wall must extend from the foundation through the roof and must be constructed so that it remains standing even if the wooden construction on one side burns completely away. The traditional fire wall is made of brick or concrete masonry, but proprietary systems using metal framing and gypsum board are also accepted in most areas.

The latter type of fire wall is especially useful in constructing attached dwellings. Building codes allow access through fire walls by means of self-closing fire doors. It should be noted here that the permitted increases in height and area relating to automatic sprinkler systems and fire walls apply to all construction systems, not just wood platform framing.

These limits are broad enough that a surprisingly high percentage of all building projects can be built using wood platform frame construction. And its economies are such that most building owners will choose it over more fire-resistant types of construction if given the opportunity. It would, however, be a mistake to assume from these facts that the national building codes are permissive when dealing with platform frame construction.

On the contrary, wood platform framing is probably regulated more closely by codes than any other construction type, including detailed specifications for framing members and fasteners and detailed drawings that show exactly how certain parts of the structure must be assembled.

Emergency exit requirements for residential structures generally include minimum size and maximum sill height dimensions for bedroom windows. An automatic fire alarm system including smoke detectors, heat detectors, or both is almost universally required in residential buildings to awaken the occupants and get them moving toward the exits before the building becomes fully involved in a fire. New building permits in many jurisdictions are requiring automatic sprinkler systems in many types of modern wood frame buildings, especially as similar, cheaper systems are developed.

Structural Design Checklist

□ Estimate the depth of rafters on the basis of the *horizontal* (not slope) distance from the outside wall of the building to the ridge board in a gable or hip roof, and the horizontal distance between supports in a shed roof. A 2 × 4 rafter spans approximately 7 ft (2.1 m), a 2 × 6 × 10 ft (3.0 m), a 2 × 8 × 14 ft (4.3 m), and a 2 × 10 × 17 ft (5.2 m).

The depth of wood light roof trusses is usually based on the desired roof pitch. A typical depth is one-quarter of the width of the building, which corresponds to a 6/12 pitch. Trusses are generally spaced 24 in (600 mm) o.c.

- □ Estimate the depth of wood floor joists as follows: 2 × 6 joists span up to 8 ft (2.4 m), 2 × 8 joists 11 ft (3.4 m), 2 × 10 joists 14 ft (4.3 m), and 2 × 12 joists 17 ft (5.2 m).
- □ Estimate the depth of wood floor trusses as one-eighteenth of their span. Typical depths of floor trusses range from 12 to 28 in (305 to 710 mm) in 2-in (51-mm) increments. Framing members in light frame buildings are usually spaced either 16 or 24 in (400 or 600 mm) o.c.

These checklist estimations are valid only for purposes of preliminary building layout and must not be used to select final member sizes. They apply to the normal range of building occupancies such as residential and commercial framing, office and institutional buildings. For manufacturing and storage buildings, use somewhat larger members.

Foundations for Light Frame Structures

Foundations for light framing, originally of stone or brick, are now made of site-cast concrete, concrete block masonry, or preservative-treated wood. Concrete and masonry foundations are highly conductive of heat and usually must be insulated to meet code requirements concerning energy conservation.

A wood foundation is easily insulated in the same manner as the frame of the house it supports. However, all foundations for light frame structures go through the standard process steps shown in Figs. 2-1 through 2-7.

Further advantages claimed for wood foundations are that they can be constructed in any weather by the same crew of carpenters that will frame the building and that they allow for easy installation of electrical wiring, plumbing, and interior finish materials in the basement. A basement of any material needs to be carefully dampproofed and drained to avoid flooding with groundwater and to prevent the buildup of water pressure that could cave in the walls.

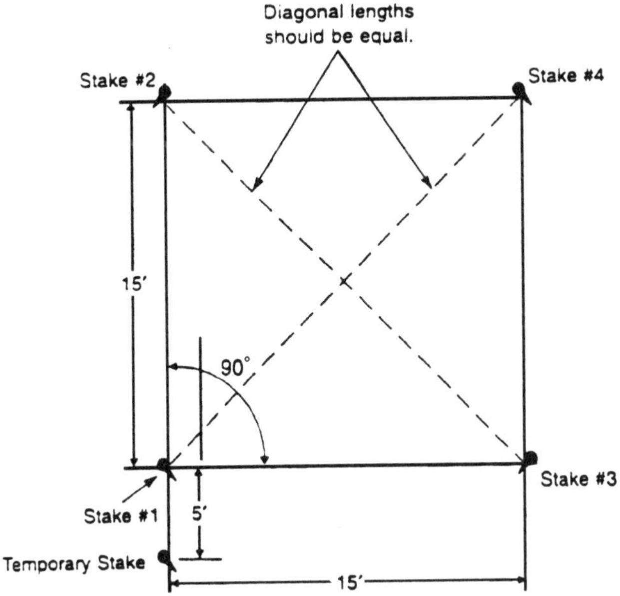

Figure 2-1 Foundation Layout

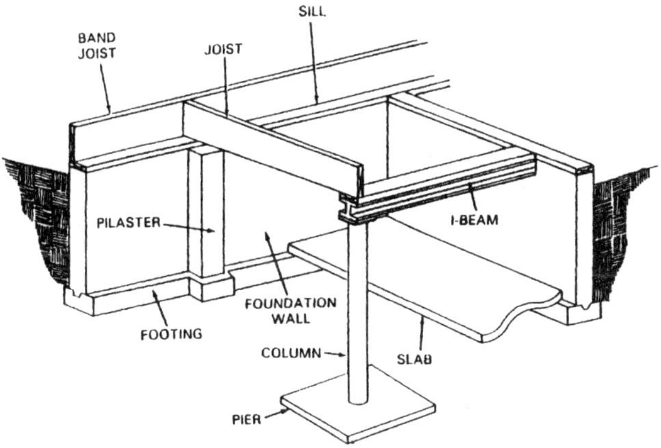

Figure 2-2 Foundation Elements

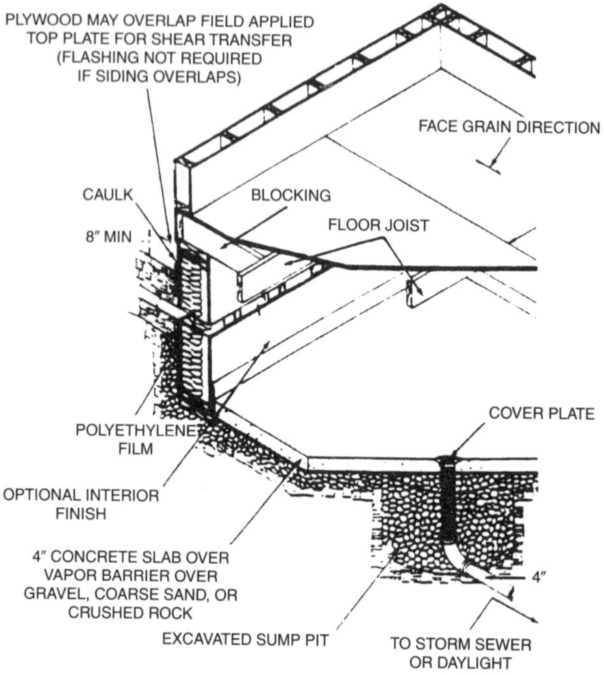

Figure 2-3 Wood Foundation Design

Planning the Framing

While it is true that an experienced carpenter can frame a simple building from the most minimal of drawings, a platform frame of wood for a custom-designed building should be planned as carefully as a frame of steel or concrete for a larger building. The architect or engineer should determine an efficient layout and the appropriate sizes for joists and rafters and communicate this information to the carpenters by means of the framing plans.

For most purposes, the member sizes can be determined using standardized structural tables. Detailed section drawings are also prepared for the major connections in the building.

The architectural floor plans serve to indicate the locations and dimensions of all the walls, partitions, and openings, and

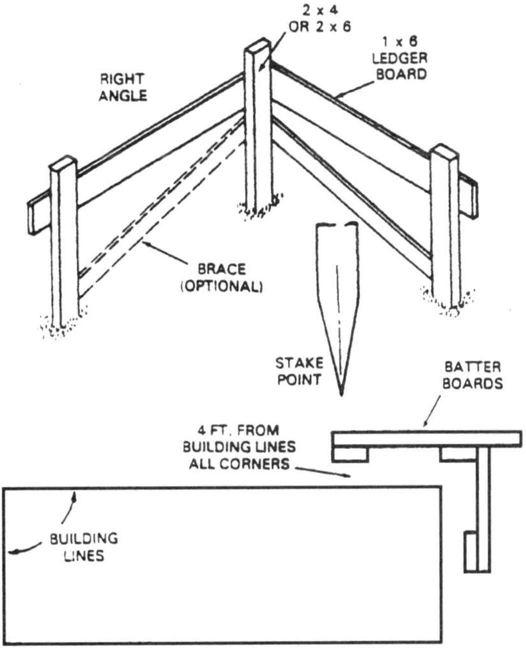

Figure 2-4 Using Batter Boards

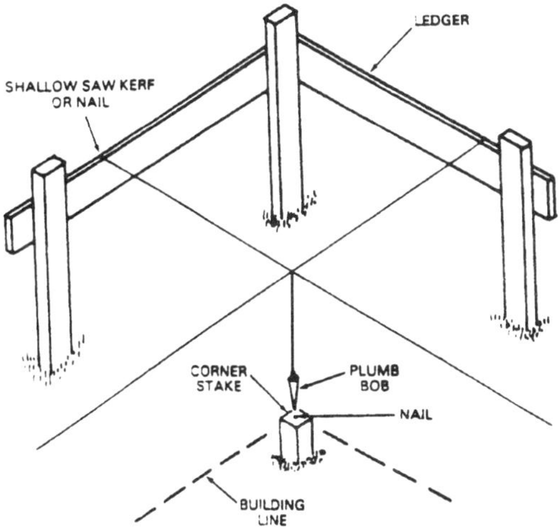

Figure 2-5 Centering Batter Boards

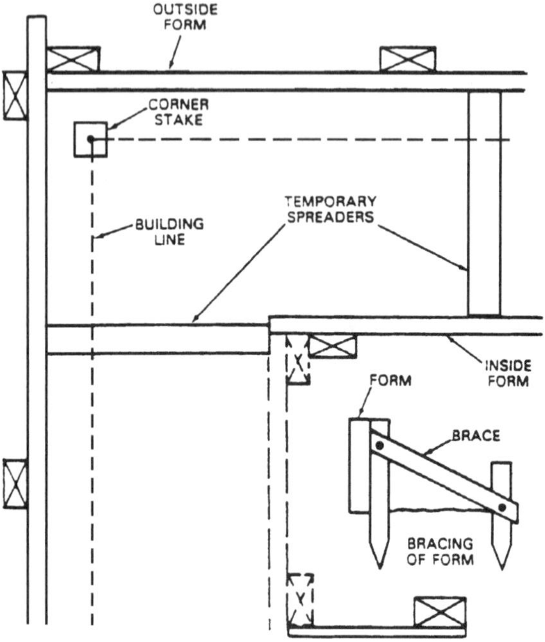

Figure 2-6 Locating Footing Forms

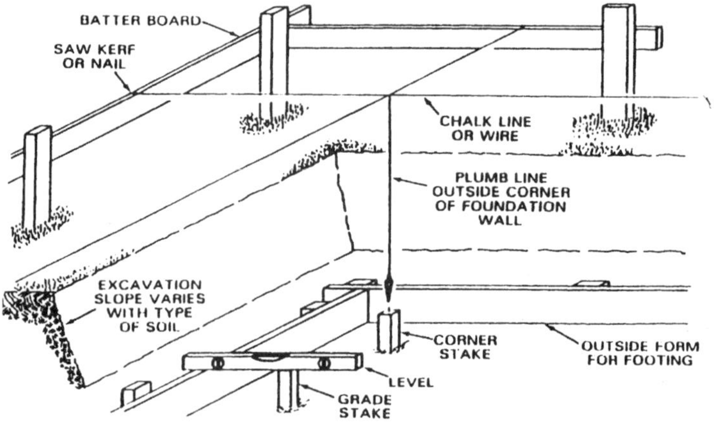

Figure 2-7 Laying Out Footing Forms

the *elevations* are drawings that show side views of the building, with vertical framing dimensions indicated as required. For most buildings, *sections* are also drawn that cut completely through the building, showing the dimensional relationships of the various floor levels and roof planes, and the slopes of the roof surfaces.

Erecting the Structural Frame

The erection of a typical platform frame (referred to, inaccurately, as "rough carpentry" in architects' specifications) can best be understood by following the sequential isometric diagrams that the plans describe. The basic simplicity of the framing process is that a platform is built, walls are assembled horizontally on the platform and tilted up into place, and another platform or a roof is built on top of the walls.

Most of the work is accomplished without the use of ladders or scaffolding, and temporary bracing is needed only to support the walls until the next level of framing is installed and sheathed.

The details of a platform frame are not left to chance. While there are countless local and personal variations in framing details and techniques, (not to mention terminology), the sizes, spacings, and connections of the members in a platform frame are closely regulated by building codes, down to the size and spacing of the studs, the size and number of nails for each connection, and the thickness of the sheathing panels.

Frame-to-Foundation Connection

The mudsill (or sill), preferably of treated or naturally decay-resistant wood, is bolted to the foundation as a base for the wood framing.

The top of the foundation is usually somewhat uneven, so the sill must be shimmed with wood shingle wedges to be able to transfer loads from the frame to the foundation. A compressible fibrous sill sealer should be inserted between the sill and the foundation to reduce air infiltration through the gap.

The normal foundation bolts are sufficient to hold most buildings on their foundations, but taller frames in areas subject to high winds or earthquakes may require more elaborate attachments.

Planning the Floor Framing

Floor framing is usually laid out in such a way that the joints of uncut subflooring panels will fall directly over joists; otherwise many panels will have to be cut, wasting both materials and time. The standard joist spacings are 16 or 24 in o.c. (on center) (406 or 610 mm), meaning that the spacing is measured to the centers of the joists. Either of these spacings automatically provides a joist at every panel joint.

Subflooring should be glued to the joists to prevent squeaking and increase floor stiffness. Plywood and oriented strand board (OSB) panels must be laid with the grain of their face plies perpendicular to the direction of the joists because these panels are considerably stiffer in this orientation. (See the section on Structural Panels.) Panels should be spaced slightly at all their edges to prevent floor buckling during construction from the expansion of storm-wetted panels.

Bridging, which is crossbracing or solid blocking between joists at midspan, is a traditional feature of floor framing. Its function is to hold the joists straight and to help them share concentrated loads. Research has shown, however, that bridging is of little use under most conditions, and many building codes no longer require it.

Where ends of joists are butted into supporting headers, as around stair openings and at changes of joist direction for projecting bays, end nails and toe nails cannot carry the full weight of the joists, and sheet-metal joist hangers must be used. Each provides a secure pocket for the end of the joist and punched holes into which a number of special short nails are driven to make a safe connection.

Planning the Wall Framing

Wall framing, like floor framing, is laid out so a framing member occurs under each joint between sheathing panels. The head carpenter initiates wall framing by laying out the stud locations on the top plate and sole plate of each wall. Other carpenters follow behind to cut the studs and headers, and assemble the walls flat on the subfloor. As each wall frame is completed, the carpenters tilt it up and nail it into position, bracing it temporarily as needed.

Sheathing of rigid panels or diagonal boards acts as permanent bracing for the walls and is applied as soon as possible after the wall is framed. Some types of sheathing panels are intended only as insulation, however, and have no structural value. Where these are used, let-in diagonal bracing is inserted into each wall before it is erected.

Corners and partition intersections must furnish nailing surfaces for each plane of exterior and interior finish materials. This requires a minimum of three studs at each intersection, unless special metal clips are used to reduce the number to two.

Framing Engineering

After each major earthquake, scientists and engineers study the performance of structures so that building codes may be evaluated and improved as needed. After the 1994 Northridge Earthquake, the City of Los Angeles and Structural Engineers of Southern California formed a joint task force to reevaluate several seismic structural provisions of the building code for wood frame buildings.

The task force recommended significant changes to the design and construction of light wood framing and specifically shear walls. These changes included lowering the height-to-width ratios and shear strength values. These changes will be appearing in future revisions of all the national building codes.

Structural wood framing systems are divided into load-bearing types, and each type is then further subdivided by the types of vertical structural elements to be used to resist the forces of inertia, horizontal distribution, and seismic load path.

Inertia is the tendency for an object at rest to remain at rest, or of an object in motion to remain in motion. Force is the energy required to move or accelerate the object. Inertia forces are forces that move or accelerate an object, and they are proportional to the object's weight. Seismic forces on structural wood frames are inertia forces that determine stress shear values, stud spacing, and wall bracing requirements as shown in Tables 2-1 through 2-3.

Table 2-1

Allowable Working Stress Shears for Wind Or Seismic Forces on Vertical Shear Walls of Fiberboard Sheathing Board[1]

Size/ Application	Fastener Type	Shear for 3-inch nail spacing around perimeter and 6-inch at intermediate studs (lb/ft of wall)
1/2 inch x 4 ft x 8 ft	No. 11 ga. galv. roofing nail, 1 1/2" long, 7/16" head	125[2]
25/32 inch x 4 ft x 8 ft	No. 11 ga. galv. roofing nail, 1 3/4" long, 7/16" head	175

1 in = 25.4 mm
1 ft = 0.305 m
1 pif = 14.5939 N/m

Notes:

1. In Category C buildings, the allowable values shall be reduced 50%.
2. The shear value may be 175 lb/ft for fiberboard classified as nail-based sheathing.

Table 2-2

Maximum Stud Spacing

Stud Size	Lateral Unsupported Stud Height (ft)	Supporting roof and ceiling only (in)	Supporting 1 floor, roof and ceiling (in)	Supporting 2 floors, roof and ceiling (in)
2x4	10	24	16	–
3x4	10	24	24	16
2x5	10	24	24	–
2x6	10	24	24	16

1 in = 25.4 mm
1 ft = 0.305 m

Table 2-3

Wall Bracing			
Building Seismic Classification	Bracing Story Location	Type of Bracing	Bracing Location
Minimum bracing for all wood frame construction where the peak velocity-related acceleration. A_V, is equal to or less than 0.05, according to 1607.	Each story	1x4 let-in diagonal bracing or in accordance with Table 2308.2.2B	Located at each corner and at every 25 ft (7620 mm) of each exterior wall.
Buildings assigned to Seismic Performance Category B and Detached one and two family dwellings (Group R3) 35 ft or less in height, assigned to category C by 1607.1	First story	In accordance with Table 2308.2.2B	25% of the length of each exterior wall.
	All other stories	In accordance with Table 2308.2.2B	Located at each corner and at every 25 ft (7620 mm) of each exterior wall.
Detached one and two family dwellings (Group R3) 35 ft or less in height, assigned to Seismic Performance Category D by 1607.1	All stories below the top story	Wood board sheathing or wood structural panels	40% of the length of each exterior wall.
	Top story	In accordance with Table 2308.2.2B	Located at each corner and at every 25 ft (7620 mm) of each exterior wall.

1 ft = 0.305 m

The effects of earthquake forces on wood framing are related to both the height and weight of the building. The higher the building's loads and weight are above the foundation, the farther the top of the building will move under the same earthquake force.

Horizontal force distribution is based on Newton's second law of motion, which states that force equals mass times acceleration. When the earthquake's sudden movement of the ground accelerates the mass of the building, horizontal forces are created and are concentrated where the weight is: the floors, roofs, and walls.

The structural framing system used must be constructed in accordance with the seismic performance category and height limitations indicated in the adopted building code. Structural framing and seismic resisting systems that are not contained in the code tables are permitted if analysis and test data are submitted that establish the dynamic characteristics and demonstrate the frame's lateral force resistance and energy absorption capacity (see Fig. 2-8) to be equivalent to the structural systems listed in the code tables.

For a dual system, the moment frame must be capable of resisting at least 25 percent of the design seismic forces. The total seismic force resistance is to be provided by the combination of the moment frame and the seismic resisting elements in proportion to their rigidities. The exception to this is that supported dual structural systems with a weight equal to or less than 10 percent of the weight of the building need not comply with this requirement.

Combinations of framing systems are different structural framing systems permitted along the two orthogonal areas of the building. A building having a combination structural system of braced frames or shear walls is limited in height where there are braced frames or shear walls that are arranged so that the braced frames or shear walls in one plane resist no more than the following proportion of the seismic design force in each direction, including torsional effects:

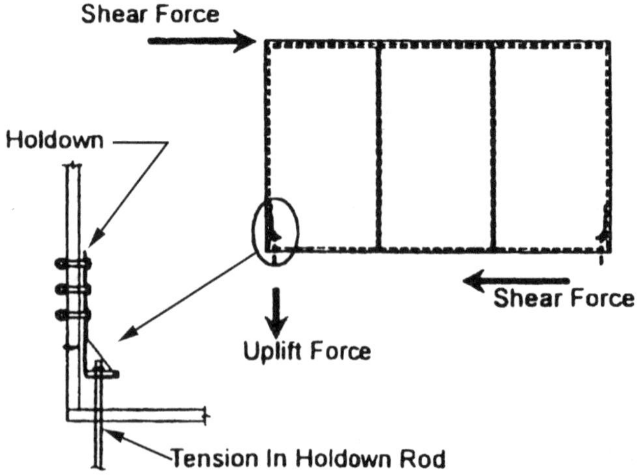

Figure 2-8 Lateral and Uplift Force

- 60 percent when the braced frame or shear walls are arranged only on the perimeter
- 40 percent when some of the braced frames or shear walls are arranged on the perimeter
- 30 percent for other structural arrangements

Moment-resisting frames that are enclosed or adjoined by more rigid elements not considered to be part of the seismic resisting system must be designed so that the action or failure of the enclosing or adjoining elements will not impair the vertical load and seismic-force-resisting capability of the frame. The structural frame design must provide for the effect of these rigid elements on the structural system at building deformations corresponding to the design story drift.

This means that the structural frame must have deformational compatibility. Every structural frame component not included in the seismic-force-resisting system in the direction under consideration must be designed to be adequate for the vertical load-carrying capacity (as shown in Fig. 2-9) and the induced moments resulting from the design story drift.

Two-story structural framing is done by the use of through-floor connectors and structural shear panels as

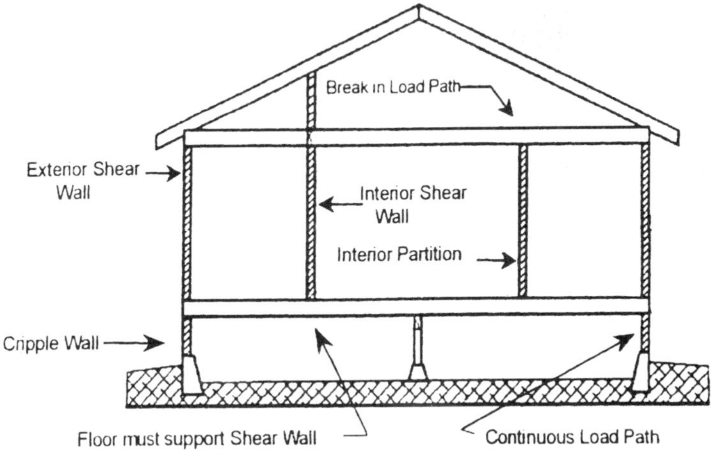

Figure 2-9 Vertical Loadbearing Capacity

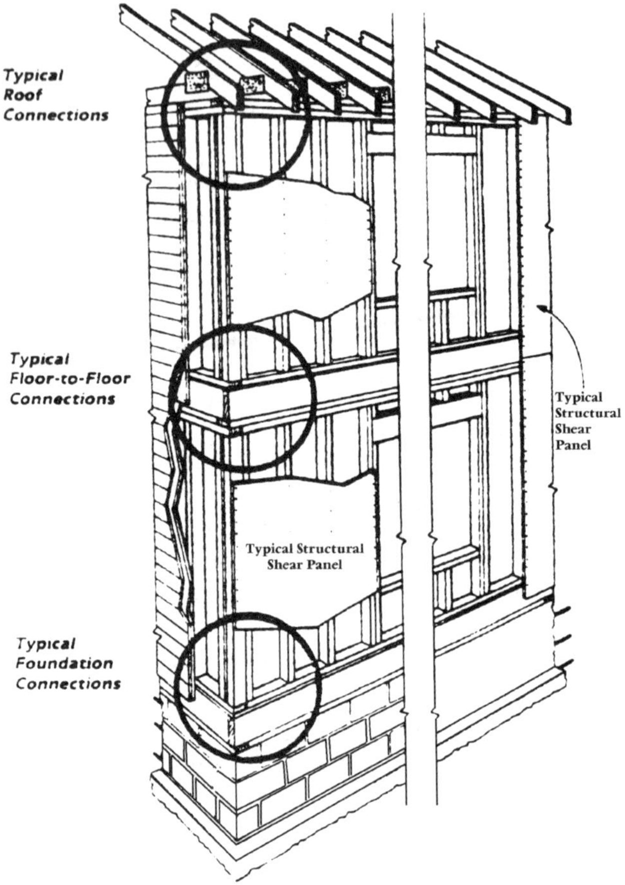

Figure 2-10 Two-story Through-floor Connections

shown in Fig. 2-10. Through-floor seismic connectors are shown in detail in Fig. 2-11.

Framing Joints

A joint is a junction, the place where two or more parts are joined. In framing, joints are often necessary to increase the length of a solid piece or to join two or more pieces at an angle. With older joints, which were bolted, keyed, or fish-plated, the members had to be much heavier than the ones that can be used with modern connectors.

Framing joints designed for supporting loads in compression are referred to as *joints-in-bearing*. Among these joints are:

Butt	Bird's-mouth
Bevel	Dovetail
Miter	Mortise and tenon
Housed	Open mortise
Notched	Step joints
Cogged	Metal shoes
Halved	

More than 100 types of joint connectors have been patented. One of the basic ideas in the design of joint connectors is the increasing of the bearing area in the joint without the need of

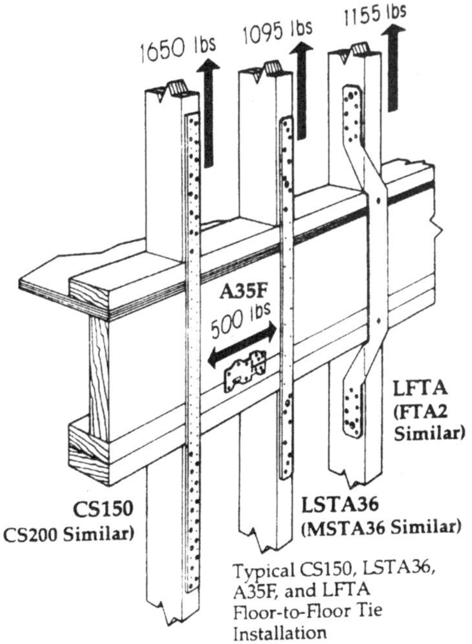

Figure 2-11 Through-floor Connectors

extensive cutting or the boring of large holes. Of the many types of connectors available, the following are among those most commonly used for structural connections in wood framing: split ring, toothed ring, claw plates, shear plates, spike grids, and clamping plates. A list of joints follows:

- **Split Rings.** These rings are manufactured in several sizes, the most commonly used being the $2^1/_2$ and 4 in. They are beveled on the inside, which allows a simple and tight-fitting installation into precut grooves. Split rings require a special cutter to cut the grooves.
- **Toothed Rings.** These rings are embedded into the faces of the members and joined by pressure. They are not well suited for the heavier types of construction.
- **Claw Plates.** The plates can be used to join a member to another member or to metal. In wood-to-wood use, the male and female units are used together. To attach wood to metal, only the female part of the connector is used.
- **Shear Plates.** These plates are set flush with the surface of the lumber in precut grooves. When joining wood to wood, they are used in pairs. When joining wood to metal, they are used as single units.
- **Spiked Grids.** Spiked grids may be flat, single-curved, or double-curved. They are used to a great extent in heavy construction.
- **Clamping Plates.** These plates hold framing members together where they cross at right angles. The *plain plate* is installed with the aid of a protecting block, while the *flanged plate* needs a protective cover plate when it is driven into position. They are usually made of sheet metal.

Joints in framing are used in relation to the manner in which they resist stresses, basically those in tension and compression. Lumber framed to resist tension by joining together by the long dimension is jointed basically in one of three ways: fishing, lapping, and scarfing.

- **Fish-Plate Joint.** A fish-plate joint may be plain or tabled. In the plain joint, there is no notching of either the lumber or the plates. The plates are bolted, lagged, or nailed directly to the two members, which are butted to each other. The fish plates may be either wood or steel.

- **Tabled Fish-Plate Joint.** In the tabled fish-plate joint, the main framing member is notched to receive the tongues, or *tables*, in the fish plates. Similar notches in the fish plates fit into tables in the other members. The number of tables in the lumber and fish plates is usually even-numbered since the holding power on both sides of the joint should be equal. If many tables are used, they must be framed very accurately; otherwise, the joint may fail because of shear in the table carrying the most load, which in turn would increase the likelihood of the remaining tables shearing.

Another factor in the use of many table joints is the fact that the notches need not be cut so deep. In general, the length of each table should be from two-thirds to the full width of the member. Shorter tables increase the tendency of shearing while longer ones may not allow bolts to hold the joint together properly.

The importance of good workmanship cannot be emphasized too strongly. Modification of the fish-plate joint will include joints where the notches and tables are indented. This type of joint requires very careful workmanship to ensure a close fit.

Notches may be made in both lumber and plates as shown in Fig. 2-12. Wedges may sometimes be used as keys; however, this type of joint is weaker than the regular fish-plate joint because (1) the key is shorter and so has a tendency to rotate under stress, and (2) the key is placed with the grain running opposite to that in the member. Since there is more "give" to the side grain of the wood in the key than the end grain in the timber, there is naturally more give to the joint.

- **Steel Fish Plates.** Tabled by having flats, called ribs, riveted to them. Fish-plate joints may be used for compression members also, as in extending the length of posts or columns where the available length of timber is inadequate. See Figs. 2-13 and 2-14.

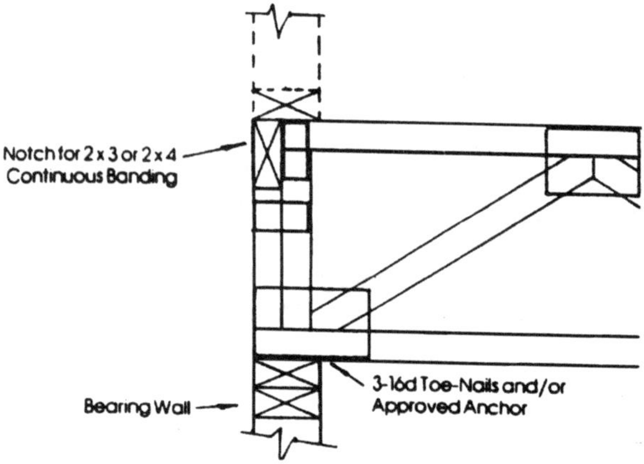

Figure 2-12 Notching

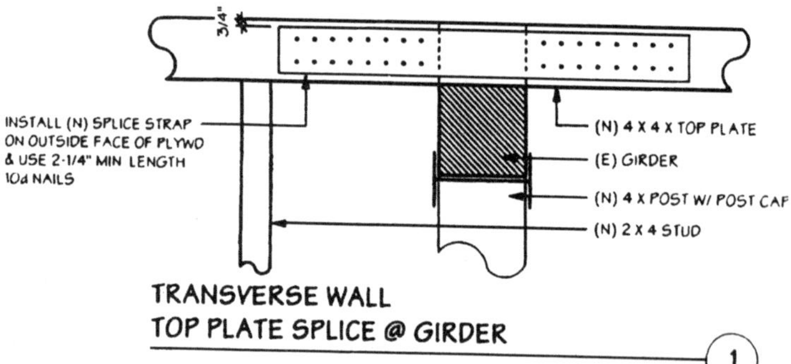

Figure 2-13 Splice Plate

- **Shear Pin Splice.** This joint is made by bolting fish-plates, or splice pads, to the main timbers to be joined, as in the plain fish-plate joint, after which holes are bored between the splice pads and the framing member. Pins are driven into these holes. Pins of solid steel were found by the University of California to be adequate to utilize the full strength of Douglas Fir. With this type of joint, it is important that the wood be well seasoned; shrinkage will result in a certain amount of slippage. Shear pins are also used in anchorage as shown in Figure 2-15.

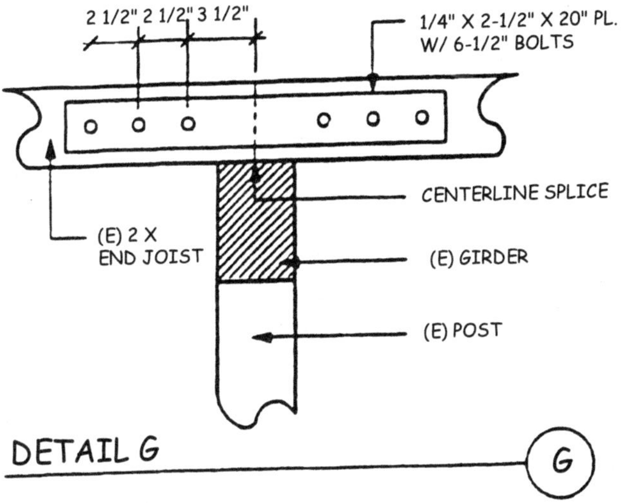

Figure 2-14 Steel Fish Plate Splice

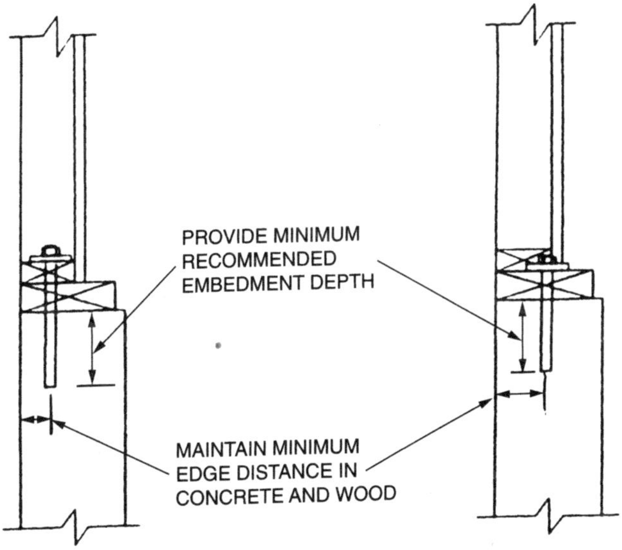

Figure 2-15 Shear Pin Anchorage

- **Lap Joint.** Typical lap joints include the simple lap joint where one member is bolted, screwed, or nailed to the other. In others, half the wood is cut away from each member to form a half-lap joint. The same joint may be notched for keys, when the joint has tension as well as compression stresses.
- **Scarf Joint.** A scarf joint is one made by fitting members together by cutting tapers or laps in each and gluing, nailing, screwing, or bolting them together.

The half-lap is perhaps the simplest scarf joint made. It may be modified by beveling the cut. The scarf joint may be tabled, or a longer notch may be cut and keys provided. Scarf joints are often used in laminated beam and truss construction.

- **Butted Joint.** The butt joint may be used in joining framing members directly end to end, end to edge, end to face, edge to edge, and face to face. Normally, there is a shear transfer that must pass through butted joints in bearing walls in critical load path locations such as the sole plate (see Figure 2-16).
- **Bevel Joint.** The bevel joint brings two pieces together at an angle. The bevel is usually the same in both pieces, though not necessarily so, and they join the members together more neatly than the butt joint.
- **Miter Joint.** To conceal the end grain, the miter joint is made by cutting two pieces to join at an angle. Usually, the pieces joined are equal in thickness or width, in which case the angle is bisected and is the same in both pieces.

There are various modifications of the miter joint in which it is combined with rabbet joints and reinforced with splines, dowels, or keys. In the rabbetted miter joint, the piece that has the shoulder should be vertical where the joint is used in bearing.

- **Housed Joint.** The housed joint is made by cutting out a section of one framing member to receive the end of another. (Some authorities, however, include as housing only those members in which the end of one framing member is tenoned and the entire width of the piece is

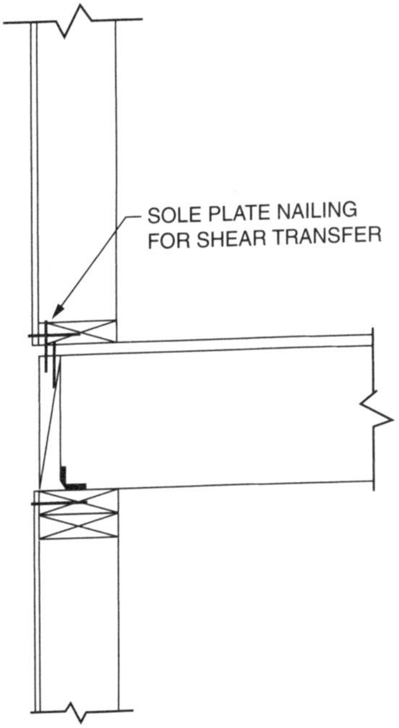

Figure 2-16 Sole Plate Butt Joint

also cut in or "housed." The same cut may also be referred to as *notching*. If both sides of the member are cut, it is referred to as double notching.

The notch is often referred to as a *gain*, and the process is called *gaining*, as in making notches for the steps in the rails of a ladder. The housing may be beveled where a framing member is let into a post. If a portion of the wood is left, the procedure is called *cogging*. This type of joint is not used much now but may be encountered.

A triangular notch cut into the edge of a timber to fit the side of another is called a *bird's mouth*. This joint occurs most frequently in framing rafters to plates. It is also used framing legs of a trestle or ledger to support the cap or top piece.

Half-lap joints made in joining two members in length is a form of scarfing. When the lap crosses the grain, it is a form of notching. The lap may be straight or beveled and should be reinforced with steel splice plates as shown in Fig. 2-17.

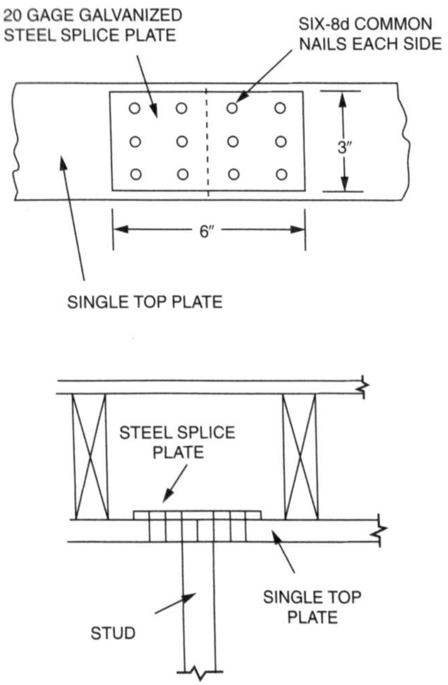

Figure 2-17 Half-lap Joints

- **Mortise and Tenon Joint.** A joint closely related to the housed joint is the *mortise and tenon*. (A tenon is a projection that fits into the mortise, or hole.) The difference is that in the mortise and tenon joint the framing member is cut on the tenoned piece to leave shoulders that bear on the face of the mortised member.

The mortise is cut a little deeper than the length of the tenon to prevent its shrinking away from the shoulders of the tenoned member, which would put all the strain on the tenon. The mortised framing member may also be notched to receive the full width or thickness of the tenoned member, in which case we have a housed tenon. These joints may be bolted or held together with wooden pins or nails.

The hole for the pin or nail is made a little closer to the shoulder of the tenon than in the corresponding mortise so that the shoulder is drawn tight to the face of the mortised timber. If made at the end of members, the joint becomes an *open* mortise and tenon.

- **Dovetail Joint.** Dovetail joints are made by cutting a wedge-shaped section out of the end of one piece and a corresponding section out of the other to permit them to fit together, yet not allowing them to be withdrawn without damage to the material. These joints resist both tension and compression. When a part of the dovetail is cut away we have what is called a *housed dovetail* or *dovetailed half-lap* joint.
- **Step Joints.** These may be single or multiple and occur in roof trusses as well as in many other kinds of framing structures. Double steps occur frequently at end joints in roof trusses. See Fig. 2-18. The ends of members at the ends of trusses are often secured by means of metal shoes. These may be forged from steel to fit specific needs. Other types of shoes are made from cast iron.

Adhesives

As a wood fastener, glue is becoming increasingly more important as continuing research and experiment add to the knowledge of adhesives and compounds. And these new designs and procedures lead to increased use of laminated products in the area of structural framing. In general, glues available on the market for woodworking are the *animal*, *synthetic resin*, *casein*, and *vegetable (starch) glues*.

Animal Glues. Animal glues are made from the hides, sinews, or bones of animals. Of these, the glues made from hides are of the highest quality while those made from bones are of the poorest. They come in different forms, usually in cake or flakes but may be ground finer for greater ease in dissolving in water. Most are made in liquid form ready to apply cold. These glues are not waterproof and should not be used where exposure to dampness is expected. Fish glues are also animal glues; generally in liquid form, they are not as strong as the glues made from cattle hides.

Where strength is less important than convenience, they are often preferred. Because animal glues are not waterproof, generally require heat to prepare, and must be clamped a short time after applying, they are not suitable for the construction industry.

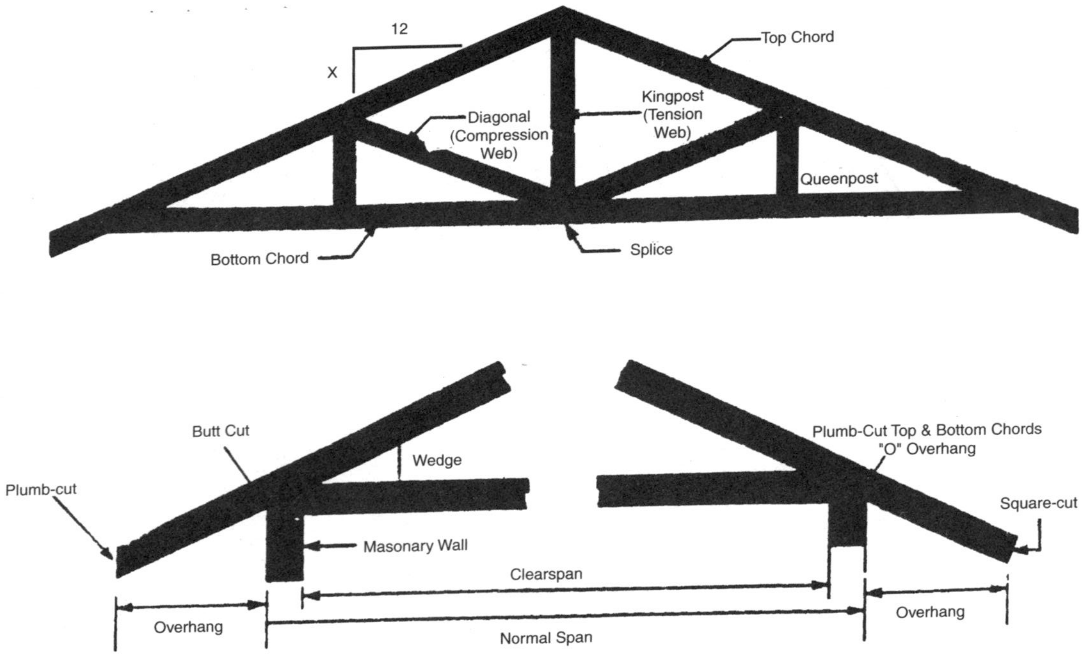

Figure 2-18 Truss Step Joints

Synthetic Resin Glues. Synthetic resin glues are basically of two types: the thermosetting and the thermoplastic. The first, once set, is not softened by heat. It is suitable for use in plywood manufacture or in laminations where hot presses are available to apply heat in setting the glues. The thermoplastic glues are suitable where the clamping and setting occurs at room temperatures. They are mixed with cold water and should be prepared in accordance with the manufacturer's recommendations.

Casein Glues. Casein glues are made from curd of milk, water, hydrated lime, and sodium hydroxide. They are mixed with cold water before use and are not waterproof. The regular glues of the casein type will stain woods such as oak and maple. A stainless type is made that can be used with these woods. It is also not waterproof.

Vegetable Glues. Vegetable glues are used mainly in the manufacture of plywood and veneered products for interior use. They have moderate moisture resistance. They should never be used for exterior work.

Epoxies and vinyls. Glues and adhesives are now made that will bond wood to concrete or metal and that can be successfully used in setting sleepers or plates on slabs that have been damp-coursed. (Damp-coursed indicates that the concrete slab has been made impervious to the seepage of water through it from below by means of a layer or layers of waterproofing materials.) The saving in time may often more than pay for the increased cost of materials. Grounds and backing may be applied by the same method.

In bonding wood to metal, some adhesives are set with heat under pressure. With others, two operations are required in completing the joint. One adhesive is applied as a primer, and after this is set on the metal, the other adhesive, a resin woodworking type, is used to fasten the wood to the primed surface of the metal. Where strength is not critical, adhesives compounded of casein and rubber latex can be used to advantage since they may be used at room temperatures.

For routine jobs, such as installing adhesives to subfloor and ceiling panels, adhesive guns may be used.

Framing Load Factors

This section describes the function of code-compliant framing load factors in relation to wood frame structural installation work. By using the word "installation," the national building codes make clear that the code regulations apply to framing systems in their manner of installation as well as in the components used.

Effective provision of wood frame structural loads and selection of component materials for particular applications involve selective engineering skills of the designer, above routine adherence to code requirements. Designers and builders must know the physical characteristics, limitations, and advantages of the many structural framing materials they use in modern construction. Safety is not automatically made a characteristic of a framing system by simply observing code compliance.

Although the building codes continually strive to attain that criterion, safety must be initially designed into any structural framing system. All residential structures must be designed and constructed to sustain, within the stress limitations specified in the code, all dead loads and all other loads. Design loads are computed from four different factors that determine either the uniform or concentrated loads, the combination of which a structural frame must bear safely. These are defined as follows:

- **Dead load** is the vertical load due to the weight of all the different permanent structural and nonstructural components of a building, such as walls, floors, roofs, and fixed service equipment
- **Live load** is the load superimposed by the use and occupancy of the building not including the wind load, earthquake load, or dead load
- **Load duration** is the period of continuous application of a given load, or the aggregate of periods of intermittent application of the same load
- **Unit live load** is a concentrated load localized in one specific area and is not distributed evenly over the general floor or loadbearing structure area

Locating and planning for HVAC and other service equipment falls under this last design load definition. Note in Fig. 2-19 how the dead load at the floor joist exceeds the computed uplift force on the joist. Residential framing structures must also be designed and constructed for roof loads to resist snow loads (if applicable to the region) and wind effects that are assumed to come from any horizontal direction.

No reduction in wind pressure is permitted by the building codes for the shielding effect of adjacent structures. Each code has a wind load formula with pressure coefficients as well as a snow load formula for its relative region that is used to calculate minimum roof live load construction.

Structural wood frame builders must also contend with three other load factors: *uplift loads*, which are negative loads imposed by winds on roofs; *hydrostatic uplift*, which is upward pressure caused by over-optimum groundwater pressure; and *earthquake loads* (see Fig. 2-20).

All of these load factors are covered by the individual national building codes with different tables relative to the regions served by the separate codes.

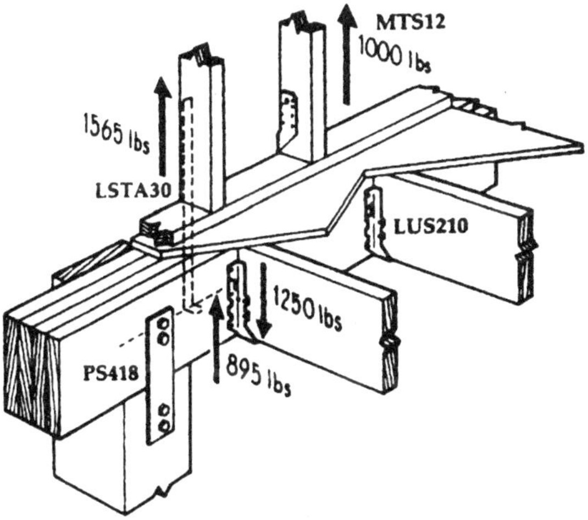

Figure 2-19 Dead Load vs. Uplift Force

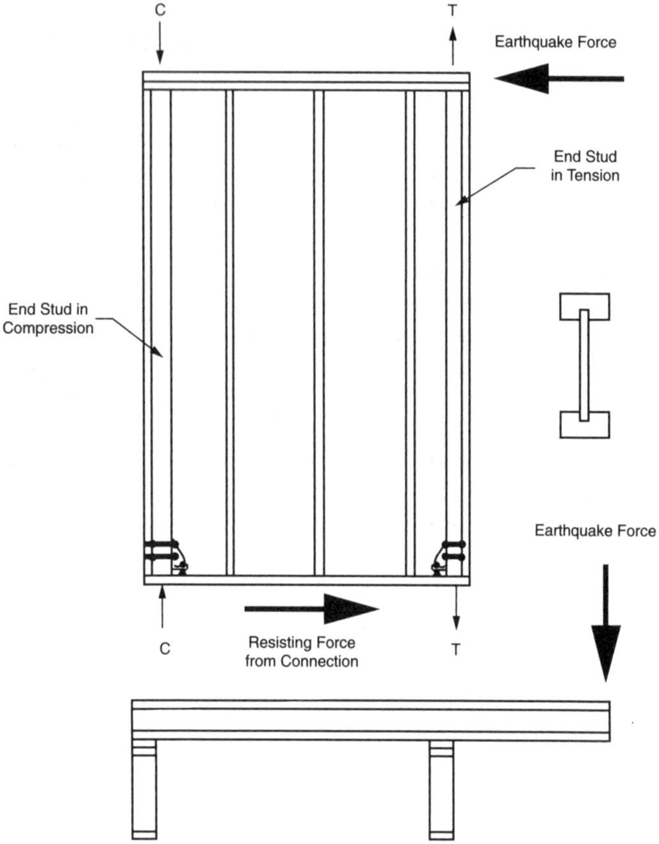

Figure 2-20 Earthquake Loads

In estimating dead loads for purposes of structural frame design, the actual weights of materials and constructions are used. The actual weight of all permanent partitions is also included in the dead load computation. Where partitions are likely to be used, although not definitely located on the plans, 20 lb/sq ft (958 Pa) is added to the dead load in the areas supporting them, except in the case of light partitioning.

The weight of fixed service equipment, such as plumbing stacks and risers, electrical feeders, and heating, ventilating, and air-conditioning systems, is also included in the floor or roof load computations as unit live loads that are concentrated wherever such equipment is supported by structural framing elements.

Uniform Floor Live Load

The uniform floor live load is the total of the greatest loads that probably will be produced by the intended use and occupancy of a residential structure. Floor live loads may be reduced in accordance with the following code provisions, per local adopted code.

Such reductions apply to slab systems designed for flexure in more than one direction, beams, girders, columns, piers, walls, and foundations. However, the following exceptions apply to this reduction variance:

- A reduction in floor live loads is not permitted in Group A occupancies
- A reduction is not permitted when the live load exceeds 100 lb/sq ft (4.8 kPa) except that the design live load for columns may be reduced 20 percent
- For live loads not exceeding 100 lb/sq ft (4.8 kPa), the design live load for any structural member supporting 150 sq ft (14 m^2) or more may be reduced at the rate of 0.08 percent per sq ft of the area supported
- Floor live load reductions must not exceed 40 percent for horizontal, 60 percent for vertical

Concentrated Floor Live Loads

In the design of floors, probable concentrated loads are taken in consideration. Where such concentrated floor loads may occur, the supporting beams, girders, and stemwalls must be designed to carry either the concentrated loads or the live load described in the national building codes section 1600, whichever produces the greater stresses.

Concentrated floor framing loads must be equal to the anticipated installed equipment, apparatus, or machinery, but cannot be less than the minimum concentrated loads specified in Table 2-4.

Interior walls, permanent partitions, and temporary partitions are designed-to resist all loads to which they are subjected but not less than 5 lb/sq ft (240 Pa) applied perpendicular to the walls, except for decorative screen walls.

Where structural framing members are arranged so as to create continuity, the distribution of the live loads, such as on adjacent spans or alternate spans, which would cause maximum design conditions must be calculated. The exception here is that all roof live loads are distributed uniformly.

The design roof live loads that take into account the effects of occupancy and water are not less than the minimum roof live loads noted in Table 2-5.

Table 2-4

Minimum Concentrated Loads[1]

Location	Load (lb)
Second floor framing (on area of 4 sq in)	2000
Finish light floor plate (on area of 1 sq in)	200
Garages	Note 2
First floors	2000
Skylight ribs and accessible ceilings	200
Stair treads (on area of 4 sq in at center of tread)	300

1 lb = 4.4482 N
1 sq in = 645.16 mm^2
1 sq ft = 0.0929 m^2

Notes:
1. Load distributed uniformly over an area of 2 $\frac{1}{2}$ sq ft unless noted otherwise.
2. Mechanical parking structures without slab or deck, passengers cars only, 1500 lb per wheel.

Table 2-5

Minimum Roof Live Load (Pounds per square foot of horizontal projection)			
	Loaded Area (sq ft) for Any Structural Member		
Roof Slope	0 to 200	201 to 600	Over 600
Flat or rise < 4 in/ft	22	18	14
Arch or dome rise $<\frac{1}{8}$ of span	20	16	12
Rise 4 in/ft to less than 12 in/ft less	20	16	12
Arch or dome rise $>\frac{1}{8}$ of span to $\frac{3}{8}$ of span	16	14	12
Rise 12 in/ft and greater	12	12	12
Awnings	5	5	5
Greenhouses	10	10	10

1 in/ft = 83.33 mm/m
1 lb/sq ft = 47.8803 Pa
1 sq ft = 0.0929 m^2

Wind Loads

All residential framing systems are designed to withstand the appropriate wind loads prescribed by local area-adjusted building code. Decreases in wind loads are not made for the effect of shielding by other structures. Wind pressures on structural framing are assumed to act normal to the surface forces considered.

Provisions of wind load framing code regulations do not apply to buildings or structures having unusual geometric shapes, response characteristics or site locations for which channeling effects or buffeting in the wake of upwind obstructions may warrant special consideration. For these cases, wind loads are based on special structural engineering calculations.

Typically, no part (component, cladding, or fastener) of a building or structure can be designed for less than 10 lb/sq ft (479 Pa). Structural members, cladding, fasteners, and systems providing for the structural integrity of the building are designed for the specified wind load velocity pres-

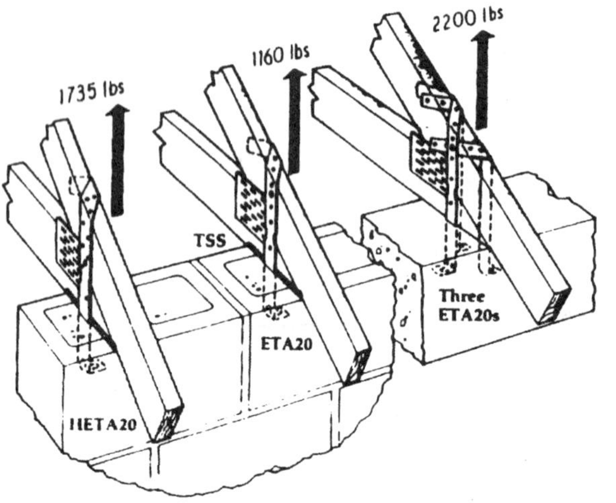

Figure 2-21 Wind Load Factors

sure. See Fig. 2-21. Coefficients include a gust factor and thus do not correspond to coefficients used in many other sources in the code tables. Mixing of coefficients and dynamic pressures from different sources is not permitted by code.

Framing members that act as both part of the main wind force resisting system and as components and cladding must be designed for separate load cases. Structural framing elements that are either directly loaded by the wind or receive wind loads originating at relatively close locations and that transfer these loads to the main wind force resisting system are subject to the wind load tables specs.

In the building code tables, what is termed the "main wind force resisting system" (MWFRS) is an assemblage of major structural framing elements assigned to provide support for secondary members and cladding. The system primarily receives wind loading from more than one surface.

Openings not accounted for are windows, doors, skylights, or other apertures in the building envelope (roof and exterior wall surfaces) that are not designed as components and cladding. The framing superstructure for a partially enclosed building must comply with all of the following conditions:

- The total area of openings in a wall or wall and roof that receives positive external pressure exceeds the sum of the areas of openings in the balance of the building envelope (walls and roof)
- The total area of openings in a wall or wall and roof that receives positive external pressure exceeds 5 percent of the area of that wall or wall and roof
- The openings in the balance of the building envelope do not exceed 20 percent

Earthquake Loads

The seismic analysis and design procedures used in the design of buildings and their structural components are based around the design seismic forces and their distribution over the height of the building. The corresponding internal forces in the structural framing components of the building are determined by the use of a linearly elastic model. Structural concepts other than these are permitted by code only when evidence is submitted showing that equivalent ductility and energy dissipation are provided.

Individual structural framing members are designed for the shears, axial forces, and moments determined in code specs. Framing connections must be designed to develop the strength of the connected members or the analysis forces, whichever is less. The design story drift of the building cannot exceed the allowable story drift requirements when the building is subjected to the design seismic forces. See Fig. 2-22.

All structural framing components of the building that transmit seismic force must be connected, with adequate strength and stiffness, through a continuous path to the final point of resistance. The foundation, in turn, must be designed to accommodate the forces developed and the movements imparted to the building by the framing system moving in design ground motions. See Fig. 2-23.

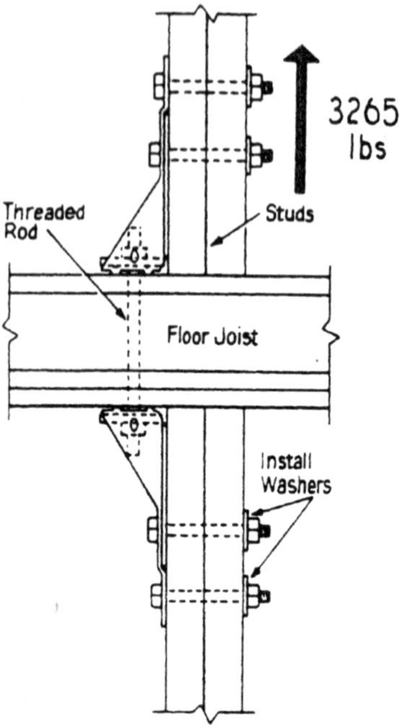

Figure 2-22 Design Seismic Force

The foundation design must consider the dynamic nature of the seismic forces on the framing system and the design ground motions that will influence it. See Fig. 2-24.

In the previous chapter you learned that a building must resist two types of forces: shear and uplift. The horizontal forces acting on the top of a shear wall not only create shear forces in the wall but will also create uplift forces at the ends of the walls. These combined with lateral force require a seismic connector that will resist in all three directions, similar to the one shown in Fig. 2-25.

Tall and narrow shear walls, in particular, have a tendency to overturn due to uplift forces created when an earthquake (lateral) force pushes the top of the wall. The shear wall's tendency to overturn may be resisted by its own weight or in combination with the weight of the structure above it.

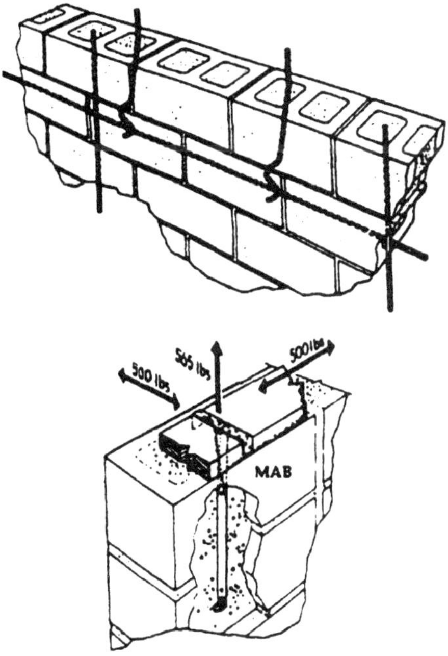

Figure 2-23 Framing Foundation Anchors

Cripple Walls

A cripple wall is a wall that is less than full story height. The cripple wall usually occurs between the first floor and the foundation (see Fig. 2-26) and is generally the weakest part of the framing of older buildings. Most jurisdictions (rightfully go) will not, however, allow less than construction-grade plywood for shear wall sheathing on a cripple wall.

A very important load path connection also must be provided at the end of each new partial cripple wall-to-foundation connection. The existing girder must be very securely connected to the new foundation or cripple wall to fully transfer the entire load along the edge of the floor that ties beyond the new partial foundation.

Where a new cripple wall is provided there are two options. The best solution is to use the existing girder as the top plate of the new cripple wall. If the girder extends as a

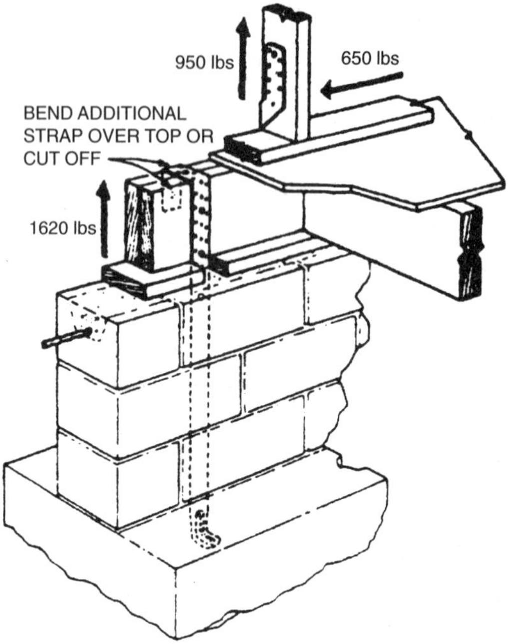

Figure 2-24 Framing to Foundation Design Forces

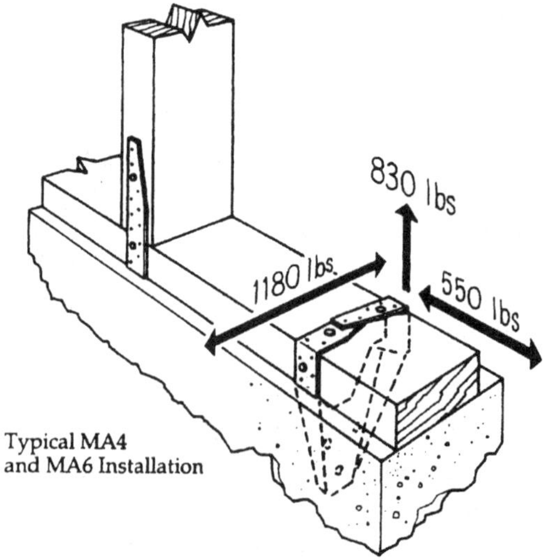

Figure 2-25 Three Directions of Force at Foundation

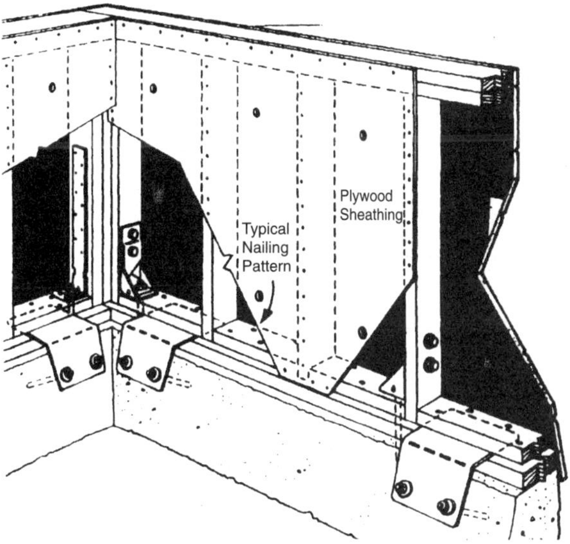

Figure 2-26 Cripple Walls

single piece the full length of the new cripple wall, no connection is necessary. However, if the girder has an existing splice location within the length of the cripple wall, a girder splice must be provided at that location.

When a new cripple wall is not needed, an anchor bolt is embedded in the foundation stem wall and secured to an L-shaped 12-gauge connector. The long leg of the connector is nailed to the girder. If using the existing girder as the cripple wall top plate is not feasible and it is cut flush with the end of the new cripple wall, a metal strap connecting the girder to the top plate of the new cripple wall must be provided as well as mudsill anchorage as shown in Fig. 2-27. Because its connection will be subject to both tension and compression forces, it is very important that the girder end be carefully cut so that it will tightly fit against the new cripple wall or foundation.

Also, to provide enough surface area to make this nailed splice connection and not interfere with the row of nailing

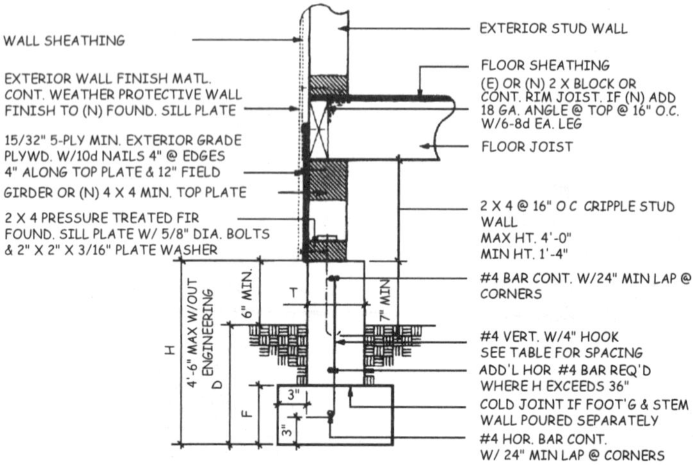

Figure 2-27 Plate to Cripple Wall Connection

along the top edge of the plywood, the new cripple wall top plate must be a nominal minimum 4 × 4 member instead of using a typical double 2 × 4 top plate.

Wood Structural Panel Sheathing

Wood structural panels include plywood, oriented strandboard (OSB), and composite panels. Wood structural panels must be of the exterior type and have a minimum thickness of $^{3}/_{8}$ in (9.5 mm), except as provided by local adopted code. All wood structural panel joints must be backed solidly with nailing pieces not less than 2 in (51 mm) wide, unless wood, wood structural panel, or particleboard sheathing is used or otherwise made waterproof as required by code. The mounting framework of wood structural panels is not required to be protected when the joints are protected by a continuous wood batt, caulking, flashing, or vertical or horizontal shiplap.

Wood structural panels when used structurally (including, among others, those used for siding, roof and wall sheathing, subflooring, diaphragms, and built-up members) must conform to the requirements for its type in U.S. Product Standards PS 1 or PS 2.

Each panel or member must be identified for grade and glue type by the trademarks of an approved testing and grading agency. Wood structural panel components must be designed and fabricated in accordance with the applicable standards listed by code and identified by the trademarks of an approved testing and inspection agency indicating conformance with the applicable standard.

Additionally, all wood structural panels when permanently exposed in outdoor applications must be of exterior type, except that wood structural panel roof sheathing exposed to the outdoors on the underside may be interior type bonded with exterior glue.

Wood structural panel diaphragms may be used to resist horizontal forces in horizontal and vertical distributing or resisting elements, provided the deflection in the plane of the diaphragm, as determined by calculations and tests, does not exceed the permissible deflection of attached distributing or resisting elements.

Permissible deflection is that deflection up to which the panel diaphragm and any attached distributing or resisting element will maintain its structural integrity under assumed load conditions, i.e., continue to support assumed loads without danger to occupants of the structure.

Connections and anchorages capable of resisting the design forces must be provided between the panel diaphragms and the resisting elements. All openings in panel diaphragms which materially affect their strength must be fully detailed on the plans and have their edges adequately reinforced to transfer all shearing stresses. The size and shape of wood structural panel diaphragms is limited. In light wood frame construction where rotation is provided for, transverse shear resisting elements normal to the longitudinal element must be provided at spacings not exceeding two times the width for particleboard diaphragms.

Horizontal and vertical panel diaphragms sheathed with particleboard may be used to resist horizontal forces not exceeding those set forth in code tables for horizontal and vertical diaphragms. The forces are dispersed through the rest of the structural frame depending on whether the floor joist are perpendicular or parallel to the bearing wall (Figs. 2-28 and 2-29). All boundary framing members must be proportioned and spliced where necessary to transmit direct stresses. Framing members have to be at least 2 in nominal in the dimension to which the particleboard is attached. In general, panel edges must bear on the framing members and butt along their centerlines.

Nails are placed not less than $^{3}/_{8}$ in (9.5 mm) from the panel edge and spaced not more than 6 in (152 mm) o.c. along panel edge bearings. Nails must be 12 in (305 mm) apart along intermediate supports and be firmly driven into the framing members. No unblocked wood structural panels less than 12 in (305 mm) wide can be used.

When either 2-in or $2^{1}/_{2}$-in fastener spacings are used with 2-in-wide framing members, the framing member adjoining panel edges must be 3 in nominal width and nails at panel edges are staggered in two lines. Framing at adjoining panel edges must be 3 in nominal or wider, and nails must be staggered where 10d nails having penetration into framing of more than $1^{5}/_{8}$ in (41 mm) are spaced 3 in (76 mm) or less o.c.

Structural sheathing must be installed in accordance with the provisions of the local adopted code when acting as wall bracing. To be considered effective as bracing, the sheathing must be at least 48 in (1219 mm) in width covering three 16-in (406 mm) stud spaces or two 24-in (610-mm) stud spaces and be fastened to the wall studs.

All vertical joints of panel sheathing must occur over studs, and all horizontal joints must occur over blocking at least equal in size to the studs. All framing in connection with sheathing used for bracing cannot be less than 2-in (51-mm) nominal thickness. Joints in lumber sheathing must occur over supports unless end matched lumber or approved clips are used, in which case each piece must bear on at least two rafters or joists.

Fiberboard sheathing, 4 × 8 ft (1219 × 2438 mm), must be applied vertically to wood studs not less than 2 in (51 mm) nominal thickness spaced 16 in (406 mm) o.c. Nailing must be in accordance with code fastener schedule tables and be

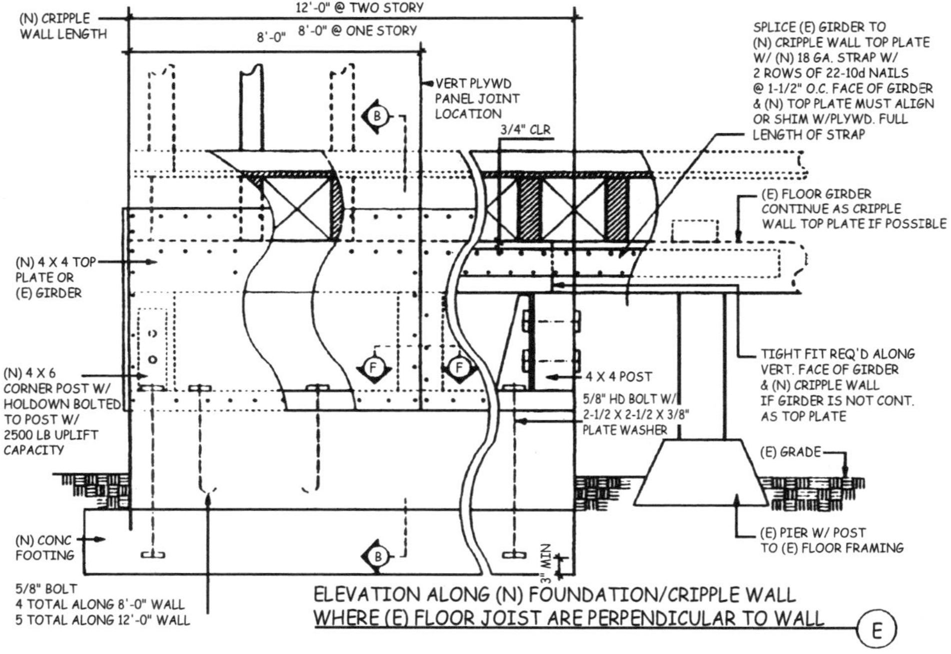

Figure 2-28 Perpendicular Diaphragm Sheathing

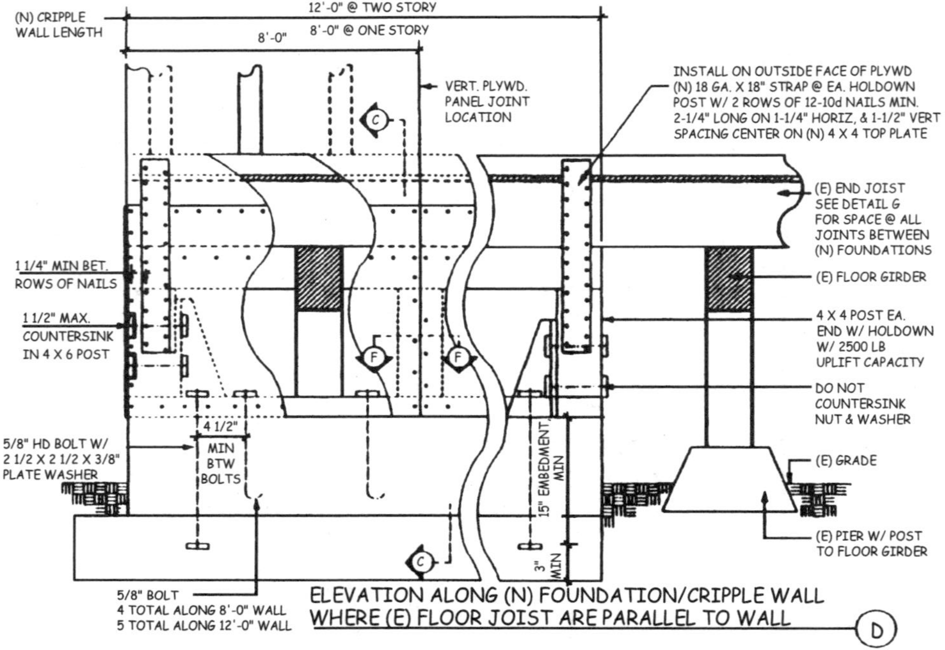

Figure 2-29 Parallel Diaphragm Sheathing

spaced not less than $^{3}/_{8}$ in (0.375 mm) from edges and ends of sheathing.

The required nail size and spacing apply to panel edges only. All wood structural panel edges must be backed with 2-in nominal or wider framing. Sheets are permitted to be installed either horizontally or vertically. For $^{3}/_{8}$-in (9.5-mm) particleboard sheets installed with the long dimension parallel to studs spaced 24 in (610 mm) o.c., nails must be spaced at 6 in (152 mm) o.c. along intermediate framing members.

For all other conditions, nails of the same size are spaced at 12 in (305 mm) o.c. along intermediate framing members. The backing framing must be 3 in (76 mm) nominal or wider and the nails staggered where nails are spaced 2 in (51 mm) o.c. or where 10d nails, having a penetration into framing of more than $1^{5}/_{8}$ in (41 mm), are used with a 3-in (76-mm) nail spacing.

The shear capacities for $^{3}/_{8}$-in (9.5-mm) and $^{7}/_{16}$-in (11.1-mm) wood structural panels applied directly to framing with 8d nails are permitted to be increased to the $^{1}/_{2}$-in (12.7-mm) shear capacities of code tables when the framing studs are spaced a maximum of 16 in (406 mm) o.c. or the panels are applied with the long dimension perpendicular to the studs.

Where particleboard is applied to both faces of a wall and the nail spacing is less than 6 in (152 mm.) o.c. on either side, panel joints must be offset to be placed on different framing members or framing must be 3 in nominal or thicker and nails on each side staggered.

Wood structural panels used for shear panels that are a part of the seismic-load-resisting system must be applied directly to the framing members. The only exceptions here are wood structural panels nailed over solid lumber decking or laminated decks. The design shear capacity is permitted to be doubled when identical materials are applied to both sides of the wall.

Where the shear capacities of the materials are not equal, the allowable shear is considered to be equal to either the shear for the side with the higher capacity or twice the shear for the side with the lower capacity.

Single diagonally sheathed lumber shear panels must be constructed of minimum 1-in-thick nominal sheathing boards laid at an angle of approximately 45° (0.785 rad) to the supports. The shear capacity for single diagonally sheathed lumber shear panels of Southern Pine or Douglas fir-larch cannot exceed 200 lb/lin ft (2.92 kN/m) of width.

Wood shear panels made up of 2-in nominal diagonal lumber sheathing fastened with 16d nails must be designed with the same shear capacities as shear panels using 1-in boards fastened with 8d nails, provided there are no splices in adjacent boards on the same support and the supports are not less than 4 in nominal depth or 3 in nominal thickness. Allowable spans are shown in Tables 2-6 through 2-8.

Double diagonally sheathed lumber shear panels must be constructed of two layers of diagonal sheathing boards laid at 90° (1.57 rad) to each other on the same face of the supporting members. Each chord is factored as a beam loaded with uniform load per foot equal to 50 percent of the unit shear due to diaphragm action. The load is assumed as act-

Table 2-6

Allowable Spans for Wood Structural Panel Wall Sheathing

Panel Span Rating	Panel thickness (in) and Construction	Maximum Stud Spacing (in) with Exterior covering nailed to: Stud	Sheathing
12/0, 16/0, 20/0 or Wall-16oc	5/16	16	16
24/0, 24/16, 32/16 or Wall-24oc	3/8, 7/16, 15/32, 1/2	24	24

1 in = 25.4 mm

Table 2-7

Allowable Spans for Particleboard Wall Sheathing Not Exposed to the Weather, Long Dimensions of the Panel Parallel or Perpendicular to Studs

		Stud Spacing (in)	
Grade	Thickness (in)	Siding Nailed to Studs	Sheathing Under Coverings Specified in 1403.6.3 and 1403.7 Parallel or Perpendicular to Studs
2-M-W	5/16	16	–
	3/8	24	16
	7/16	24	24
2-M-1 and 2-M-2	3/8	16	–
	1/2	16	16

1 in = 25.4 mm

Table 2-8

Minimum Thickness of Wall Sheathing

Sheathing Type	Minimum thickness (inches)	Maximum stud spacing (inches)
Wood boards	5/8	24
Fiberboard	7/16	16
Wood Structural Panel	According to Table 2308.1B	
2-M-1 or 2-M-W Particleboard	According to Table 2308.1C	
Gypsum sheathing	1/2	16
Gypsum wallboard	1/2	24
Reinforced cement mortar	1	24

1 in = 25.4 mm

ing normal to the chord in the plane of the diaphragm in either direction.

The span of the chord or any portion thereof is the distance between framing members of the diaphragm such as the joists, studs, and blocking that serve to transfer the assumed load to the sheathing. The shear capacity of double diagonally sheathed diaphragms of Southern Pine or Douglas fir-larch cannot exceed 600 lb/lin ft (8.76 kN/m) of width. The shear capacity is adjusted by reduction factors of 0.82 for Group III framing species and 0.65 for Group IV framing species.

Shear panels must be constructed of wood structural panels, manufactured with exterior glue, not less than 4 × 8 ft (1.22 × 2.44 in), except at boundaries and changes in framing. Wood structural panels must be designed to resist shear only, and chords, collector members, and boundary members must be designed to transfer the axial forces. Boundary members must be connected at all corners. Wood structural panels less than 12 in (305 mm) wide must be blocked.

The design shear capacity of particleboard panels must be in accordance with code section 2300 for diaphragms and section 2311 for shear walls.

Shear panels must be constructed with particleboard sheets not less than 4 × 8 ft (1.22 × 2.44 mm), except at boundaries and changes in framing.

Particleboard panels must be designed to resist shear only, and chords, collector members, and boundary members must be connected at all corners. Particleboard panels less than 12 in (305 mm) wide must be blocked.

Shear panels on a wood stud wall sheathed with other materials such as lath and plaster, gypsum sheathing board, gypsum wallboard, or fiberboard sheathing must be built to resist all three directional forces previously detailed. Nails must be spaced at least $^3/_8$ in (9.5 mm) from the edges and ends of boards and panels. The maximum height-to-width ratio is 1:5. The shear capacities used in design cannot be cumulative with the shear capacities of other materials applied to the same wall.

Notes

Notes

Chapter

3

Fire-resistant Framing

Code Compliance

Structural framing is first intended to resist the three forces of engineering dynamics: compression, tension, and shear, as detailed previously. Second, it is important for the framer to understand that the building codes are intended only to assure that construction is done with end-user safety in mind, i.e., to provide a system that is essentially free from hazard for the public.

The fire-, life-, safety-protection of the people who will inhabit your structure should always be your bottom-line criterion as a structural framer, just as it is with the building codes. All construction is rated by potential hazard to people using the building or structure. Fire is the number one life-threatening hazard in all buildings, and this chapter, fire resistance, is paramount to all modern light wood framing.

Fire-resistance requirements of the framing structure are based on fire-resistance ratings. Materials, thicknesses, and assemblies that have successfully performed under tests made by a recognized laboratory in accordance with the requirements of ASTM E119 or based on calculations and accepted engineering practice in the local adopted code are accepted for specific ratings. The exception here is that in determining the fire-resistance rating of exterior bearing walls, compliance with the ASTM E119 criteria for unexposed surface temperature rise and ignition due to passage of flame or gases is required only for a period of time corresponding to the required fire-resistance rating of an exterior nonbearing wall with the same horizontal separation distance. This maximum flame-spread allowable is regulated by the code tables as shown in Table 3-1.

The fire resistance of building components such as the structural framing, interior and exterior walls, nonload-bearing partition walls, roofs, floors, exterior doors, and windows determines the type of construction. If any of these primary building components is reduced in type of construction, the entire building is automatically classed in a lower classification.

If one of these elements were reduced, a Type I building with required 3-hour structural frame, 2-hour floors, and 4-hour exterior bearing walls would be lowered in classification to Type V due to the lack of full compliance with Type I construction requirements.

Types of construction are classified by two factors, noncombustible and fire-resistive, that need to be defined here. Noncombustible materials will not burn, but they will transmit heat in a fire and contribute to flame spread, which maintains combustion if the fire becomes hot enough. Noncombustible materials include steel, masonry, iron, and concrete. However, any material that is exposed to enough heat will melt and disintegrate structurally. Fire-resistive materials will burn when exposed to prolonged flame. The building codes define these two properties as follows:

"*Noncombustible*, as applied to a building construction material means a material that, in the form in which it is used, is either one of the following:

1. Material of which no part will ignite and burn when subjected to fire. Any material approved as conforming to

Table 3-1

FLAME-SPREAD CLASSIFICATION

MATERIAL QUALIFIED BY:	
Class	Flame-spread index
I	0-25
II	26-75
III	76-200

MAXIMUM FLAME-SPREAD CLASS

OCCUPANCY GROUP	ENCLOSED VERTICAL EXITWAYS	OTHER EXITWAYS	ROOMS OR AREAS
A	I	II	II
E	I	II	III
I	I	I	II
H	I	II	III
B, F, M and S	I	II	III
R-1	I	II	III
R-3	III	III	III
U	NO RESTRICTIONS		

this definition is considered noncombustible within the meaning of the building codes.

2. Material having a structural base of noncombustible material as defined in Item 1 above, with a surfacing material not over $^1/_8$ inch thick that has a flame-spread rating of 50 or less as approved by the Fire Code.

Noncombustible does not apply to surface finish materials. Material required to be noncombustible for reduced clearances to flues, heating appliances, or other sources of high temperature must refer to material conforming to Item 1 above. No material must be classed as noncombustible which is subject to increase in combustibility or flame-spread rating, beyond the limits herein established, through the effects of age, moisture, or other atmospheric conditions."

It should be noted here that there is a difference between flame-spread rating and fire-resistive. A flame-spread rating compares the time it takes flame to spread on the surface of the tested material against the time it takes for the same amount of flame to spread on untreated, dried red oak of uniform density. On the flame-spread rating scale, red oak is given a rating of 100 representing maximum flame spread and at the other end of the scale is cement asbestos board, which is given a rating of 0 representing minimum flame spread.

Fire resistance is determined by an Underwriter's Laboratories (UL) test of the material. A constant flame is applied to one side of the material in a controlled laboratory test environment, and the temperature change (delta factor) is measured on the other side. If, in the test time allotment, the exposed side does not exceed 250°F, the material meets specifications for a UL listed flame resistance and receives an according UL flame resistance rating. Interior and exterior wall coverings as well as door and window assemblies are all sample tested and rated for fire resistance in this manner. These ratings are classified by the different types of construction as shown in Table 3-2.

Table 3-2

TYPES OF CONSRTUCTION–FIRE-RESISTIVE REQUIREMENTS (In Hours)
For details, see occupancy section in Chapter 3, type of construction sections in this chapter and sections referenced in this table.

BUILDING ELEMENT	TYPE I	TYPE II			TYPE III		TYPE IV
	Noncombustible				Combustible		
	Fire-resistive	Fire-resistive	1-Hr.	N	1-Hr.	N	H.T.
1. Bearing walls–exterior	4 Sec. 602.3.1	4 Sec. 603.3.1	1	N	4 Sec. 604.3.1	4 Sec. 604.3.1	4 Sec. 605.3.1
2. Bearing walls–interior	3	2	1	N	1	N	1
3. Nonbearing walls–interior	4 Sec. 602.3.1	4 Sec. 603.3.1	1 Sec. 603.3.1	N	4 Sec. 604.3.1	4 Sec. 604.3.1	4 Sec. 605.3.1
4. Structural frame[1]	3	2	1	N	1	N	1 or H.T.
5. Partitions–permanent	1	1	1	N	1	N	1 or H.T.
6. Shaft enclosures[3]	2	2	1	1	1	1	1
7. Floors and floor-ceilings	2	2	1	N	1	N	H.T.
8. Roofs and roof–ceilings	2 Sec. 602.5	1 Sec. 603.5	1 Sec. 603.5	N	1	N	H.T.
9. Exterior doors and windows	Sec. 602.3.2	Sec. 603.3.2	Sec. 603.3.2	Sec. 603.3.2	Sec. 604.3.2	Sec. 604.3.2	Sec. 605.3.2
10. Stairway construction	Sec. 602.4	Sec. 603.4	Sec. 603.4	Sec. 603.4	Sec. 604.4	Sec. 604.4	Sec. 605.4

N– No general requirements for fire resistance. H.T.– Heavy Timber

[1]Structural frame elements in an exterior wall that is located where openings are not permitted or where protection of openings is required, shall be fire exposure as required for exterior bearing walls or the structural frame, whichever greater.

[2]Fire-retardant-treated wood (see Section 207) may be used in the assembly, provided fire-resistace requirements are maintained.

[3]For special provisions, see Section 304.6, 306.6 and 711.

To meet the framing fire-resistance requirements of the building code, it must be determined that materials, constructions, and assemblies of all framing systems have successfully performed under accepted tests that are prescribed in the local adopted building code.

Fire-resistance Framing Terms and Definitions

Annular Space. The opening around a penetrating item.

Damper, Fire. A damper arranged to seal off airflow automatically through part of an air duct system inside the framing structure, so as to restrict the passage of heat.

Door Assembly, Fire. A combination of the fire door, frame, hardware, and other accessories that together provide a specific degree of fire protection to the opening.

Door, Fire. A door and its assembly, so constructed and assembled in place as to give protection against the passage of fire.

Draft Stopping. Materials and framing components installed to prevent the movement of air, smoke, gases, and flame to other areas of the building through large concealed passages, such as attic spaces and floor assemblies with suspended ceilings or open-web trusses.

Fire Area. The aggregate floor area enclosed and bounded by fire walls, exterior walls, or fire separation assemblies of a building.

Fireblocking. Framing components and materials installed to prevent the movement of flame, smoke, and gases to other areas of a building through small concealed passages in building components such as floors, walls, and stairs.

Fire Partition. A vertical assembly of framing having protected openings and designed to restrict the spread of fire.

Fire-protection Rating. The time in hours, or fractions thereof, that an opening protective assembly will resist fire exposure as determined by the test standard specified in the Fire Code.

Fire-resistance. That property of framing assemblies and materials that prevents or retards the passage of excessive heat, hot gases, or flames under conditions of use.

Fire-resistance Rating. The time in hours or fractions thereof that framing assemblies or materials will resist fire exposure as determined by the fire test specified in the Fire Code.

Fire-resistive Joint System. An assemblage of specific framing components, materials, or products that are designed, tested, and fire resistance rated in accordance with ASTM E119 to resist (for a prescribed period of time) the spread of fire through joints made in or between fire-resistance rated framing assemblies.

Fire Separation Assembly. A horizontal or vertical fire-resistance rated framing assembly having protected openings, and designed to restrict the spread of fire.

Fire Separation Distance. The distance in feet measured from the building face to the closest interior lot line, to the centerline of a street or public way or to an imaginary line between two buildings on the same property.

Fire Wall. A 4-hour fire-resistant wall, having protective openings, which restricts the spread of fire and extends continuously from the foundation to or through the roof, with sufficient structural stability under fire conditions to allow collapse of construction on either side without collapse of the wall.

Fire Window. A window constructed and glazed to give protection against the passage of fire.

Joint. The linear opening in or between adjacent fire-resistance rated assemblies.

Party Wall. A fire wall on an interior lot line used or adapted for joint service between two buildings.

Protected Construction. That in which all structural members are constructed, chemically treated, covered, or protected so that the individual unit or the combined assemblage of all such units has the required fire-resistance rating specified for its particular application and includes protected combustible and protected noncombustible construction.

Self-Closing. As applied to a fire door or other opening protective means normally closed and equipped with an approved device that will ensure closing after having been opened for use.

Shaft. An enclosed space extending through one or more stories of a building, connected vertical openings in successive floors, or floors and the roof.

Single-membrane Penetration. An opening through a single membrane (one side) of a fire-resistance rated wall, roof-and-ceiling or floor-and-ceiling assembly made to accommodate electrical, mechanical, plumbing, environmental, and/or communication systems.

Smoke Barrier. A continuous membrane that will resist the movement of smoke.

Smoke Compartment. A space within a building enclosed by smoke barriers or fire-separation assemblies on all sides, including top and bottom.

Splice. The result of a factory or field method of joining or connecting two or more lengths of a fire-resistive joint system into a continuous assembly.

Through Penetration. Specific building materials or assemblies of materials that are designed and installed to prevent the spread of fire through openings that are made in fire-resistance rated floors and walls to accommodate through penetrations for electrical, mechanical, plumbing, environmental, and/or communication systems. The *F* rating indicates the period of time that the through penetration firestop system is capable of preventing the passage of flame to the unexposed (nonfire) side of the assembly in conjunction with an acceptable hose stream test performance.

The *T* rating indicates the period of time that the through-penetration firestop system is capable of preventing the passage of flame and a maximum individual temperature rise of 325°F (163°C) above ambient temperature on the unexposed (nonfire) side of the assembly in conjunction with acceptable hose stream test performance.

Vertical Opening. An opening through a floor or roof.

Wall: Fire Separation Wall. A fire-resistance rated assembly of materials having protected openings that is designed to restrict the spread of fire.

The period of time a structural frame maintains the ability to confine a fire or continues to perform a given structural function or both is critical. Any wood product that, when impregnated with fire-resistant chemicals by pressure process or other means during manufacture, should have, when tested in accordance with ASTM E84, a flame-spread index of 25 or less and show no evidence of significant progressive combustion when the test is continued for an additional 20 min. In addition, the flame front should not progress more than 10 ft 6 in (3200 mm) beyond the centerline of the burners at any time during the test.

All model building codes contain extensive requirements pertaining to fire- and life-safety considerations (see Table 3-3). These provisions involve all aspects of the building design but are strongly influenced by the type of structural materials used, i.e., combustible or noncombustible. The following sections briefly discuss the fire- and life-safety issues most commonly associated with light wood framing and roof framing with wood-steel composite trusses.

Type of Construction

Generally, the building codes of North America permit wood trusses in heavy timber, ordinary, and frame construction classifications. Worth noting here is the regulation that wood trusses may only be used in the heavy timber classification when all members of the truss meet the minimum size requirement of that classification.

Ordinary and frame constructions are typically subdivided into protected (fire-resistive) and nonprotected categories. Because of the minimum member size required, heavy timber construction is assumed to have a fire endurance in keeping with the occupancies, building size, etc., allowed by the building codes. See Table 3-3.

Table 3-3

RATED FIRE-RESISTIVE PERIODS FOR VARIOUS WALLS AND PARTITIONS

ITEM NUMBER	CONSTRUCTION	MINIMUM FINISHED THICKNESS FACE-TO-FACE (inches) × 25.4 for mm			
		4 Hr.	3 Hr.	2 Hr.	1 Hr.
1-1.1	Solid units (at least 75 percent solid).	8		6	4
1-2.1	Hollow brick units (at least 71 percent solid).		8		
1-2.2	Hollow brick units (at least 60 percent solid, cells filled with loose-fill insulation).	8			
1-2.3	Hollow brick units (at least 64 percent solid).	12			
1-2.4	Hollow brick, not filled	5.0	4.3	3.4	2.3
1-2.5	Hollow brick unit wall, grout or filled with perlite vermiculite or expanded shale aggregate.	6.6	5.5	4.4	3.0
1-3.1	Hollow (rowlock).	12		8	
1-4.1	Cavity wall consisting of two 3" (76 mm) (actual) solid clay brick units separated by 2" (51 mm) air space, joint reinforcement every 16" (406 mm) on center vertically.		8		
1-4.2	Cavity wall consisting of two 4" (100 mm) nominal solid clay brick units separated by 2" (51 mm) air space, $^{1}/_{4}$" (6.4 mm) metal ties for 3 sq. feet (0.28m^2) of wall area.	10			
1-5.1	4" (100 mm) nominal thick units at least 75 percent solid backed with a hat-shaped metal furring channel $^{3}/_{4}$" (19 mm) thick formed from 0.021" (0.53 mm) sheet metal attached to the brick wall on 24" (610 mm) centers with approved fasteners, and $^{1}/_{2}$" (12.7 mm) Type X gypsum wallboard attached to the metal furring strips with 1" (2.5 mm) long Type S screws spaced 8" (203 mm) on center.			5	

Table 3-3 (*Continued*)

ITEM NUMBER	INSULATING MATERIAL USED	× 25.4 for mm			
		4 Hr.	3 Hr.	2 Hr.	1 Hr.
1-5.1	Vermiculite concrete, 1:4 mix by volume over paperbacked wire fabric lath wrapped directly around column with additional 2" by 2" (51 mm by 51 mm) 0.065 inch/0.065 inch (1.65 mm/1.65 mm) (No. 16/16 B.W. gage) wire fabric placed $^3/_4$" (19 mm) from outer concrete surface. Wire fabric tied with 0.049 inch (1.24 mm) (No. 18 B.W. gage) wire spaced 6" on center for inner layer and 2" (51 mm) on center for outer layer.	2			
1-6.1	Perlite or vermiculite gypsum plaster over metal lath wrapped around column and furred $1^1/4$" (32 mm) from column flanges. Sheets lapped at ends and tied 6" (153mm) intervals with 0.049 inch (1.24 mm) (No. 18 B.W. gage) tie wire. Plaster pushed through to flanges.	$1^1/_2$	1		
1-6.2	Perlite or vermiculite gypsum plaster over self-furring metal lath wrapped directly around column, lapped 1" (25 mm) and tied at 6" (153 mm) intervals with 0.049 inch (1.24 mm) (No. 18 B.W. gage) wire.	$1^3/_4$	$1^3/_8$	1	
1-6.3	Perlite or vermiculite gypsum plaster on metal lath applied $^3/_4$" (19 mm) cold-rolled channels spaced 24 inches (610 mm) apart vertically and wrapped flatwise around column.	$1^1/_2$			
1-6.4	Perlite or vermiculite gypsum plaster over 2 layers of $^1/_2$" (12.7 mm) plain full length gypsum lath applied tight to column flanges. Lath wrapped with 1" (25.4 mm) hexagonal mesh of No. 20 gage wire and tied with doubled 0.035 inch diameter (0.89 mm) (No. 18 B.W. gage) wire ties spaced 23" (584 mm) on center. For three-coat work the plaster mix for the second coat shall not exceed 100 pounds (45.4 kg) of gypsum to $2^1/2$ cubic feet (0.07 m^3) of aggregate for the three-hour system.	$2^1/_2$	2		
1-6.5	Perlite or vermiculite gypsum plaster over one layer of $^1/_2$" (127 mm) plain full-length gypsum lath applied tight to column flanges. Lath tied with doubled0.049 inch (1.24 mm) (No. 18 gage) wire ties spaced 23" (584 mm) on center and scratch coat wrapped with 1" (25 mm) hexagonal mesh 0.035 inch (0.89 mm) (No.20 B.W. gage) wire fabric. For three coat work, the plaster mix for the second coat shall not exceed 100 pounds (45.4 kg) of gypsum to $2^1/_2$ cubic feet (0.07 m^3) of aggregate.		2		
1-7.1	Multiple layers of $^1/_2$" (12.7 mm) gypsum wallboard adhesively secured to column flanges and successive layers. Wallboard applied without horizontal joints. Corner edges of each layer staggered. Wallboard layer below outer layer secured to column with doubled 0.049 inch (1.24 mm) (No. 18 B.W. gage) steel wire ties spaced 15" (381 mm) on center. Exposed corners taped and treated.			2	1

Protected or fire-resistive construction, when required for a particular building, requires the trusses to be part of an assembly having the specified fire-endurance rating. Fire-endurance assembly ratings are established by laboratory test. Tests on individual manufacturers pertain only to the specific truss or trusses of that manufacturer.

Numerous similar tested assemblies are listed in the fire-resistive assemblies publications of Underwriters Laboratories, Inc. Others may be found in the manufacturer's literature and research reports published by various building code authorities.

Also unless properly fire- and draft-stopped, the truss space located between the structural decking and the ceiling construction can act as a horizontal chimney with consequent rapid spread of flame and smoke. Adherence to building code requirements and careful detailing are necessary to ensure adequate fire and draft stopping.

Sprinkler Systems and Wood Truss Design

The addition of an approved automatic sprinkler system to a building permits modification to be made to the basic fire provision of the building codes. Allowable areas and heights of buildings may be increased and in some codes fire endurance requirements may be reduced. In wood truss construction incorporating a ceiling, it is generally necessary to install sprinkler heads both in the concealed truss space and below the ceiling construction.

When supported by the trusses, sprinkler systems can impose substantial structural loads. While small branch lines usually can be accounted for as additional uniformly applied dead load, main lines may require reduced truss spacing or additional trusses to support these loads. Code tables give approximate wet-line weights for sprinkler systems depending on the diameter of the pipe as well as standpipe requirements. See Table 3-4.

Spacing of sprinkler line supports varies with diameter. The resulting concentrated loads can be a controlling criterion in the design of lightweight prefabricated trusses.

Table 3-4

STANDPIPE REQUIREMENTS

OCCUPANCY	NONSPRINKLERED BUILDING		SPRINKLERED BUILDING	
× 304.8 for mm × 0.0929 for m^2	Standpipe Class	Hose Requirement	Standpipe Class	Hose Requirement
1. Occupancies exceeding 150 ft. in height and more than one story	III	Yes	I	No
2. Occupancies 4 stories or more but less than 150 ft. in height, except Group R, Division 3	I and II (or III)	Yes	I	No
3. Group A Occupancies with occupant load exceeding 1,000	II	Yes	No requirement	No
4. Group A , Division 2.1 Occupancies over 5,000 square feet in area used for exhibition	II	Yes	II	Yes
4. Groups I; H; B; S; M; F, Division 1 Occupancies less than 4 stories in height but greater than 20,000 square feet per floor	II	Yes	No requirement	No
6. Stages more than 1,000 square feet in area	II	No	III	No

Connections must be capable of resisting certain horizontal forces as well as vertical loads.

Moreover, all connectors must be of an approved type. Generally, the horizontal forces and approved connections are those given in Standard NFPA 13 National Fire Protection Association.

Some lightweight wood truss manufacturers can provide recommendations and approved details particularly suited to their products. In any case, sprinkler loads and connections must be considered for their effect on wood truss design.

In considering the structural effects of a sprinkler system, a frequent problem is not having the final sprinkler system layout at the time of the structural design. However, preliminary layouts and maximum line sizes usually can be obtained and appropriate allowances provided in the wood truss design.

Unfortunately, a number of contractors and designers have suffered the embarrassing and costly situation (not only with wood trusses but with other structural framing systems) of having a structural system in place before considering sprinkler system loads. It is essential that these loads and their connections be considered before final truss layout and design.

Fire-Retardant-Treated (FRT) Wood Trusses

The use of FRT wood in trusses generally reduces fire insurance rates. Some codes allow limited use of such members in noncombustible construction types. However, the fire-retardant treatments cause a significant reduction in the allowable strength properties of wood, and in certain cases the chemicals used in this process can cause corrosion of the metal connectors. Allowable design values, including connector design values, are available from the company providing the treatment and redrying service.

It is particularly important to provide for adequate ventilation in roof systems using FRT wood trusses combined with high-temperature environments. Some truss manufacturers do not recommend the use of fire-retardant treatments with their products.

Fire-blocking barriers are installed to resist the movement of flame and gases to other areas of a building through small concealed passages in building components such as floors, walls, and stairs. This cuts down on the flame-spread over a surface.

When insulation or other materials that may change the capacity for heat dissipation are added to or subtracted from fire-resistant roof or ceiling framing assemblies whose fire ratings are listed in the code or listed in the plans or reference documents, fire test results or other substantiating data must be submitted to the building official to show that the required fire resistance time period is not reduced.

Thicknesses established by fire tests are established as *minimum* requirements for fire-resistance only, and do not preclude the application of other requirements of the building code where consideration of strength, durability, or stability require greater thicknesses.

Where support columns require a fire-resistance rating, the entire column, including its connections to beams or girders, must be protected. Where the column extends through a ceiling, fire resistance of the column must be continuous from the top of the floor through the ceiling space to the underside of the floor deck above.

Lintels over openings in walls must also be protected to provide a fire resistance rating at least equal to that required for beams, except that fire resistance may be omitted from the bottom flange of lintels, shelf angles, and plates not a part of the structural frame or with a span of 6 ft (1829 mm) or less. In 1-hour fire-resistant construction, the ceiling may be omitted over unusable crawl space and flooring may be omitted when unusable attic space occurs above.

Fire-resistant Framing Separations

All construction materials must bear identification showing the fire performance rating of that material. Such identification is issued and attached or marked on the material by the approved agency having a service for inspection of materials at the factory.

Allowable openings in fire-rated walls are permitted for small openings in fire-rated walls or ceilings conduit, piping, duct work, and electrical boxes made of ferrous metal. Maximum openings up to 100 sq in within 100 sq ft of ceiling are permitted. However, fire-resistive floors must be continuous. Floor and roof penetrations as openings for electrical and mechanical equipment must be enclosed. The exception here is that vent and service piping may be installed through fire-resistive floors as long as the assembly does not reduce the required fire-resistance rating of the floor. Conduit and pipes cannot be embedded in structural members or required fire-protective coverings.

Protective coverings for penetrations in fire-rated framing assemblies must meet certain thickness requirements that are specified in Chapter 7 in each of the national building codes. These thickness requirements apply to all construction and materials used in fire-resistive framing assemblies. These minimum thicknesses are measured in the material's net thickness and do not include any open or hollow spaces behind the protection.

Interior walls and partitions that enclose vertical openings such as stairways are required to have a fire resistance rating that extends from floor to floor or from floor to roof. These walls must be continuous through all concealed spaces such as the space above a suspended ceiling. The framing structure must have a fire-resistance rating equal to or greater than the fire-resistance rating required for the vertical enclosure. Where the openings are offset at intermediate floors, the offset and floor construction must have a fire resistance of not less than that required for the enclosing partitions.

All other partition walls required to have a fire-resistance rating must extend from the top of the floor below to the ceiling above and be securely attached as part of the framing structure. Where the ceiling is not a part of a framing assembly having a fire-resistance rating at least equal to that required for the partition wall, the partition must be constructed tight against the floor or roof deck above. The construction of the partitions or ceilings and any openings must be such as to prevent the spread of smoke.

In multiresidential framing such as apartments or townhouses, each unit is considered a separate building and must be separated from adjoining units by a fire-rated party wall or by the use of separate exterior walls meeting the requirements for zero clearance from property lines as required for the type of construction. Separate exterior walls include one of the following:

1. A parapet not less than 18 in (457 mm) above the roof line.
2. Roof sheathing of noncombustible material or fire-retardant-treated wood, for not less than a 4-ft (1219-mm) width on each side of the exterior dividing wall.
3. One layer of $^5/_8$ in (15.9 mm) Type X gypsum board attached to the underside of roof decking, for not less than a 4-ft (1219-mm) width on each side of the exterior dividing wall.

When less than three stories in height, townhouses may be separated by a single wall meeting the following requirements:

1. Such wall provides not less than a 2-hour fire-resistance rating. Plumbing, piping, vent ducts, electrical systems, or other building services cannot be installed within or through the 2-hour wall, unless such materials and methods of penetration have been tested in accordance with adopted code specifications.
2. Such wall is continuous from the foundation to the underside of the roof sheathing or has a parapet extending not less than 18 in (457 mm) above the roof line. When such wall terminates at the underside of the roof sheathing, the roof sheathing for not less than a 4-ft (1219-mm) width on each side of the wall must be of a noncombustible material, fire-retardant-treated wood, or

one layer of $^5/_8$-in (15.9-mm) Type X gypsum wallboard attached to the underside of the roof decking.

3. Each dwelling unit sharing such a wall is designed and constructed to maintain its structural integrity independent of the unit on the opposite side of the wall. Said wall may be penetrated by roof and floor structural members provided that the fire-resistance rating and the structural integrity of the wall is maintained.

Party walls and fire walls cannot extend more than 3 ft (914 mm) above the roof. However, fire walls are not required to extend above the roof where the roof is rated noncombustible in Types I, II, and IV construction, or noncombustible or fire-retardant treated wood for an area within 40 ft (12.2 in) of each side of the wall in Types III, V, and VI framing construction.

Party walls and fire walls cannot be framed to extend less than 18 in (457 mm) past exterior intersecting walls of combustible construction or exterior noncombustible walls with combustible projections or veneers. The party or fire wall also must extend not less than 18 in (457 mm) past any combustible projection or veneer. Party walls or fire walls must extend to the inside facing of the exterior surface of noncombustible construction.

Fireblocking

Framing fireblocking involves constructing a system to prevent, for a prescribed time period, the passage of flame, heat, and hot gases through openings that penetrate an entire fire-resistant framing assembly in order to accommodate conduits, tubing, pipes, or similar items.

Fireblocking must be provided in all walls and partitions to cut off all concealed draft openings (both horizontal and vertical) and to form a fire barrier between floors and between the upper floor and the roof space. This is such a critical area of safety in structural framing that fireblocking cannot be covered or concealed until inspected by the building official.

Walls and stud partitions must be fireblocked at floors, ceilings, and roofs. Fireblocking in noncombustible partitions is not required at the ceiling for suspended ceiling systems. Fireblocking must consist solely of approved noncombustible materials unless otherwise specified in code and all component materials must be securely fastened in place.

Any openings between the edge of a floor deck and an exterior wall must be sealed using an approved material or assembly of materials designed and tested for this purpose. The material must remain in place, sealing the opening, for a time period at least equal to the required fire-resistance rating of the floor deck. Enclosed attic and floor spaces formed of combustible construction must also be divided.

Openings in the wall framing for steel electrical outlet boxes not exceeding 16 sq in (0.010 m^2) are permitted provided the area of such openings does not aggregate more than 100 sq in (0.06 m^2) for any 100 sq ft (9 m^2) of fire-resistant wall area or shaft-enclosure wall area.

Outlet boxes on opposite sides of the fire-resistant wall or shaft enclosure must be separated by a horizontal distance of not less than 24 in (610 mm). However, openings for electrical outlet boxes of any material are permitted provided such boxes are listed for use in fire-resistant assemblies and are installed in accordance with their listings.

Ducts may penetrate fire-resistant walls provided the duct penetrations are protected with approved fire dampers. Fire dampers used for protection of duct penetrations must also be installed in accordance with their listing. A typical stud cavity firestop is shown in Fig. 3-1. Other lateral framing firestops are shown intersecting the floor joists in the framing details in Fig. 3-2.

All firestopping is estimated by takeoffs from the exterior walls plan view combined with the interior walls plan view. Often they are combined on the prints in one view similar to Fig. 3-3.

The minimum thickness of wall insulating material for fire-resistive periods is regulated by the structural parts to be protected. These differ by the way the wall is constructed such as an exterior or interior wall, wood studs

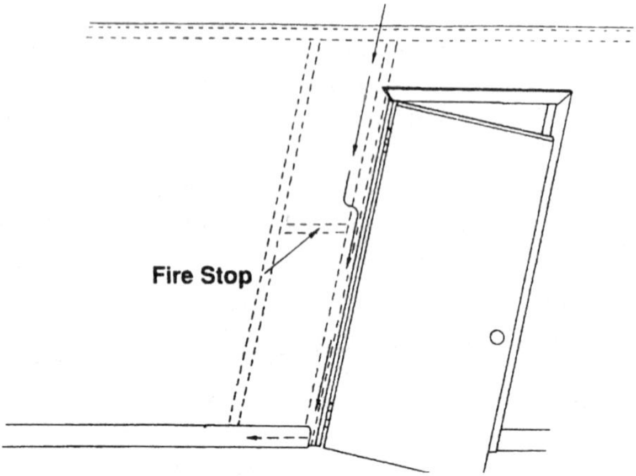

Figure 3-1 Fire Stop

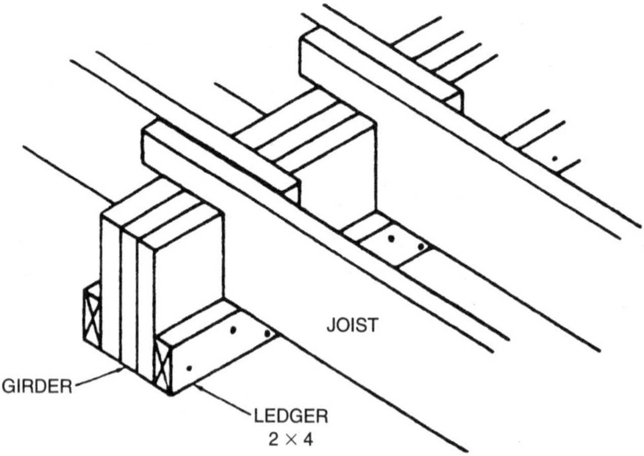

Figure 3-2 Fire Stops

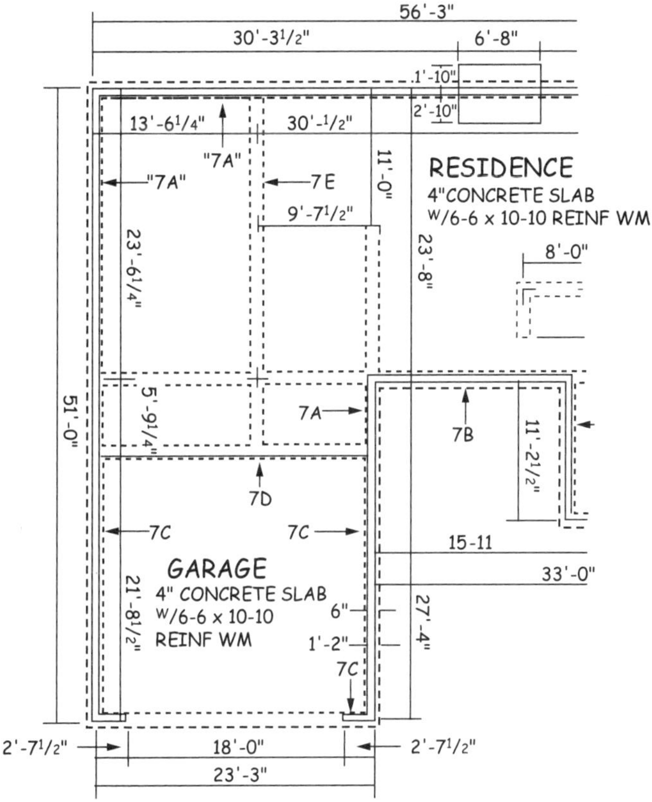

Figure 3-3 Plan View

interior partition with plaster on each side, noncombustible studs interior partition with gypsum wallboard on each side, or wood studs interior partition with gypsum wallboard on each side.

The wall's fire-resistive ratings are further broken down in classifications of wall construction and roof types as shown in the following tables. Tables 3-5 shows fire-resistive ratings by the components of the wall's construction. Tables 3-6 through 3-8 show cross-referenced fire ratings by floor or roof construction.

Table 3-5

ITEM NUMBER	CONSTRUCTION	MINIMUM FINISHED THICKNESS FACE-TO-FACE (inches) × 25.4 for mm			
		4 Hr.	3 Hr.	2 Hr.	1 Hr.
2-1.1	Two cells in wall thickness, units at least 40 percent solid.				8
2-1.2	Two cells in wall thickness, units at least 43 percent solid.				
2-1.3	Two cells in wall thickness, units at least 46 percent solid.				8
2-1.4	Two cells in wall thickness, units at least 49 percent solid.			8	
2-1.5	Three or four cells in wall thickness, units at least 40 percent solid.				8
2-1.6	Three or four cells in wall thickness, units at least 43 percent solid.			8	
2-1.7	Three or four cells in wall thickness, units at least 48 percent solid.			8	
2-1.8	Three or four cells in wall thickness, units at least 53 percent solid.		8		
2-1.9	Three cells in wall thickness, units at least 40 percent solid.			12	
2-1.10	Three cells in wall thickness, units at least 45 percent solid.		12		
2-1.11	Three cells in wall thickness, units at least 49 percent solid.		12		
2-1.12	Two units and three or four cells in wall thickness, units at least 40 percent solid.		12		
2-1.13	Two units and three or four cells in wall thickness, units at least 45 percent solid.	12			
2-1.14	Two units and three or four cells in wall thickness, units at least 53 percent solid.	12			
2-1.15	Two or three units and four or five cells in wall thickness, units at least 40 percent solid.	16			

ITEM NUMBER	CONSTRUCTION	MINIMUM FINISHED THICKNESS FACE-TO-FACE (inches) × 25.4 for mm			
		4 Hr.	3 Hr.	2 Hr.	1 Hr.
3-1.1	One cell in wall thickness, units at least 40 percent solid.				4
3-1.2	One cell in wall thickness, units at least 30 percent solid.			6	
3-1.3	Two cells in wall thickness, units at least 45 percent solid.				6
3-1.4	One cell in wall thickness, units at least 40 percent solid.				4
3-1.5	One cell in wall thickness, units at least 30 percent solid.			6	
4-1.1	Two cells in wall thickness, units at least 40 percent solid.				8
4-1.2	Two cells in wall thickness, units at least 49 percent solid.			8	
4-1.3	Three or four cells in wall thickness, units at least 53 percent solid.		8		
4-1.4	Two cells in wall thickness, units at least 46 percent solid.				8
4-1.5	Three cells in wall thickness, units at least 40 percent solid.			12	
4-1.6	Two units and three cells in wall thickness, units at least 40 percent solid.		12		
4-1.7	Two units and three or four cells in wall thickness, units at least 45 percent solid.	12			
4-1.8	Three cells in wall thickness, units at least 45 percent solid.		12		
4-1.9	Three cells in wall thickness, units at least 49 percent solid.		12		

Table 3-5 (*Continued*)

ITEM NUMBER	CONSTRUCTION	MINIMUM FINISHED THICKNESS FACE-TO-FACE (inches) × 25.4 for mm			
		4 Hr.	3 Hr.	2 Hr.	1 Hr.
8-1.7	One 4" (100 mm) unit cored 30 percent maximum with $^3/_4$"(19 mm) vermiculite gypsum plaster on one side.			$4^1/_2$	
8-1.8	One 4" (100 mm) unit cored 39 percent maximum with $^3/_4$"(19 mm) gypsum plaster on one side.				$4^1/_2$
9-1.1	$^3/_4$" (19 mm) by 0.055 inch (1.4 mm) (No. 16 carbon sheet steel gage) vertical cold-rolled channels, 16"(406 mm) on center with 2.5 pound (1.13 kg) flat metal lath applied to one face and tied with 0.049 inch (1.24 mm) (No. 18 B.W. gage) wire at 6" (153 mm) spacing. Gypsum plaster each side mixed 1:2 by weight, gypsum to sand aggregate.				2
9-1.2	$^3/_4$" (19 mm) by 0.055 inch (1.4 mm) (No. 16 carbon sheet steel gage) cold-rolled channels 16" (406 mm) on center with metal lath applied to one face and tied with 0.049 inch (1.24 mm) (No. 18 B.W. gage) wire at 6" (153 mm) spacing. Perlite or vermiculite gypsum plaster each side. For three-coat work, the plaster mix for the second cost shall not exceed 100 pounds (45.4 kg) of gypsum to $2^1/_2$ cubic feet (0.071m^3) of aggregate for the one-hour system.			$2^1/_2$	2
9-1.3	$^3/_4$" (19 mm) by 0.055 inch (1.4 mm) (No. 16 carbon sheet steel gage) vertical cold-rolled channels, 16" (406 mm) on center with $^3/_8$" (9.5 mm) gypsum lath applied to one face and attached with sheet metal clips. Gypsum plaster each side mixed 1:2 by weight, gypsum to sand aggregate.				2

ITEM NUMBER	CONSTRUCTION	MINIMUM FINISHED THICKNESS FACE-TO-FACE (inches)			
		× 25.4 for mm			
		4 Hr.	3 Hr.	2 Hr.	1 Hr.
9-2.1	Studless with $^{1}/_{2}$" (12.7 mm) full-length plain gypsum lath and gypsum plaster each side. Plaster mixed 1:1 for scratch coat and 1:2 for brown coat, by weight, gypsum to sand aggregate.				2
9-2.2	Studless with $^{1}/_{2}$" (12.7 mm) full-length plain gypsum lath and perlite or vermiculite gypsum plaster each side.			$2^{1}/_{2}$	2
9-2.3	Studless partition with $^{3}/_{8}$" (9.5 mm) rib metal lath installed vertically, adjacent edges tied 6" (153 mm) on center with No. 18 gage wire ties, gypsum plaster each side mixed 1:2 by weight, gypsum to sand aggregate.				2
10-1.1	Perlite mixed in the ratio of 3 cubic feet (0.085 m^3) to 100 pounds (45.4 kg) of portland cement and machine applied to stud side of$1^{1}/_{2}$" (38 mm) mesh by 0.058 inch (1.47 mm) (No. 17 B.W. gage) paper-backed woven wire fabric lath wire-tied to 4" (102 mm) deep steel trussed wire studs 16" (406 mm) on center. Wire ties of 0.049 inch (1.24 mm) (No. 18 B.W. gage) galvanized steel wire 6" (153 mm) on center vertically.			$3^{1}/_{8}$	
11-1.1	$^{3}/_{4}$" (19 mm) by 0.055 inch (1.4 mm) (No. 16 carbon sheet steel gage) cold-rolled channels, 12" (305 mm) on center with 2.5-pound (1.13 kg) flat metal lath applied to one face and tied with 0.049 inch (1.24 mm) (No. 18 B.W. gage) wire at 6" (153 mm) spacing. Neat gypsum plaster applied each side.			2	
12-1.1	One full-length layer $^{1}/_{2}$" (12.7 mm) Type X gypsum wallboard laminated to each side of 1" (25.4 mm) full-length V-edge gypsum coreboard with approved laminating compound. Vertical joints of face layer and coreboard staggered at least 3" (76 mm).			2	

Table 3-5 (*Continued*)

ITEM NUMBER	CONSTRUCTION	MINIMUM FINISHED THICKNESS FACE-TO-FACE (inches)			
		× 25.4 for mm			
		4 Hr.	3 Hr.	2 Hr.	1 Hr.
17-1.4	2" by 4" (51 mm by 102 mm) fire-retardant-treated wood studs spaced 24" (610 mm) on center with one layer of $^5/_8$" (15.9 mm) thick Type X gypsum wallboard applied with face paper grain (long dimension) parallel to studs. Wallboard attached with 6d cooler or wallboard nails at 7" (178 mm) on center.				$4^3/_4$
17-1.5	2" by 4" (51 mm by 102 mm) wood studs 16" (406 mm) on center with two layers of $^5/_8$" (15.9 mm) Type X gypsum wallboard each side. Base layers applied vertically and nailed with 6d cooler or wallboard nails at 9" (229 mm) on center. Face layer applied vertically or horizontally and nailed with 8d cooler or wallboard nails at 7" (178 mm) on center. For nail-adhesive application, base layers are nailed 6" (153 mm) on center. Face layers applied with coating of approved wallboard adhesive and nailed 12" (305 mm) on center.			6	
17-1.6	2" by 3" (51 mm by 76 mm) fire-retardant-treated wood studs spaced 24" (610 mm) on center with one layer of $^5/_8$" (15.9 mm) thick Type X gypsum wallboard applied with face paper grain (long dimension) at right angles to studs. Wallboard attached with 6d cement-coated box nails spaced 7" (178 mm) on center.				$3^5/_8$

ITEM NUMBER	CONSTRUCTION	MINIMUM FINISHED THICKNESS FACE-TO-FACE (inches) × 25.4 for mm			
		4 Hr.	3 Hr.	2 Hr.	1 Hr.
18-1.1	Exterior surface with $^3/_4$" (19 mm) drop siding $^1/_2$" (12.7 mm) gypsum sheathing on 2" by 4"(51 mm by 102 mm) wood studs at 16" (406 mm) on center; interior surface treatment as required for one-hour-rated exterior or interior 2" by 4" (51 mm by 102 mm) wood stud partitions. Gypsum sheathing nailed with $1^3/_4$" (44.5 mm) by No. 11 gage by $^7/_{16}$" (11.1 mm) head galvanized nails at 8"(203 mm) on center. Siding nailed with 7d galvanized smooth box nails.				Varies
18-1.2	2" by 4" (51 mm by 102 mm) wood studs at 16" (406 mm) on center with metal lath and $^3/_4$" (19 mm) cement plaster on each side. Lath attached with 6d common nails 7" (178 mm) on center driven to 1" (25 mm) minimum penetration and bent over. Plaster mix 1:4 for scratch coat and 1:5 for brown coat, by volume, cement to sand.				$5^3/_8$
18-1.3	2" by 4" (51 mm by 102 mm) wood studs at 16" (406 mm) on center with $^7/_8$" (22 mm) cement plaster (measured from the face of the studs) on the exterior surface with interior surface treatment as required for interior wood stud partitions in this table. Plaster mix 1:4 for scratch coat and 1:5 for brown coat, by volume, cement to sand.				Varies
18-1.4	$3^5/_8$" (92 mm) No. 16 gage noncombustible studs 16" (406 mm) on center with $^7/_8$" (22 mm) cement plaster (measured from the face of the studs) on the exterior surface with interior surface treatment as required for interior, nonbearing, noncombustible stud partitions in this table. Plaster mix 1:4 for scratch coat and 1:5 for brown coat, by volume, cement to sand.				Varies

Table 3-5 *(Continued)*

ITEM NUMBER	CONSTRUCTION	MINIMUM FINISHED THICKNESS FACE-TO-FACE (inches) × 25.4 for mm			
		4 Hr.	3 Hr.	2 Hr.	1 Hr.
18-1.5	$2^1/_4$" by $3^3/_4$" (57 mm by 95 mm) clay face brick with cored holes over $^1/_2$" (12.7 mm) gypsum sheathing on exterior surface of 2" by 4" (51 mm by 102 mm) wood studs at 16" (406 mm) on center and two layers $^5/_8$" (15.9 mm) Type X gypsum wallboard on interior surface. Sheathing placed horizontally or vertically with vertical joints over studs nailed 6" (153 mm) on center with $1^3/_4$" (44.5 mm) by No. 11 gage by $^7/_{16}$" (11.1 mm) head galvanized nails. Inner layer of wallboard placed horizontally or vertically and nailed 8" (203 mm) on center with 6d cooler or wallboard nails. Outer layer of wallboard placed horizontally or vertically and nailed 8" (203 mm) on center with 8d cooler or wallboard nails. All joints staggered with vertical joints over studs. Outer layer joints taped and finished with compound. Nail heads covered with joint compound. 0.035 inch (0.91 mm) (No. 20 galvanized sheet gage) corrugated galvanized steel wall ties $^3/_4$" by $6^5/_8$" (19 mm by 198 mm) attached to each stud with two 8d cooler or wallboard nails every sixth course of bricks.			10	
18-1.6	2" by 6"(51 mm by 153 mm) fire-retardant-treated wood studs 16" (406 mm) on center. Interior face has two layers of $^5/_8$" (15.9 mm) Type X gypsum wallboard with the base layer placed vertically and attached with 6d box nails 12"(305 mm) on center. The face layer is placed horizontally and attached with 8d box nails 8" (203) on center at joints and 12"(305 mm) on center elsewhere. The exterior face has a base layer of $^5/_8$" (15.9 mm) Type X gypsum wallboard placed vertically with 6d box nails 8" (203 mm) on center at joints and 12"(305 mm) on center elsewhere. An approved building paper is next applied, followed by self-furred exterior lath attached with $2^1/_2$" (63.5 mm) No. 12 gage galvanized roofing nails with a $^3/_8$" (9.5 mm) diameter head and spaced 6" (153 mm) on center along each stud. Cement plaster consisting of a $^1/_2$" (13 mm) brown coat is then applied. The scratch coat is mixed in the proportion of 1:3 by weight, cement to sand with 10 pounds (4.54 kg) of hydrated lime and 3 pounds (1.36 kg) of approved additives or admixtures per sack of cement. The brown coat is mixed in the proportion of 1:4 by weight cement to sand with the same amounts of hydrated lime and approved additives or admixtures used in the scratch coat.			$8^1/_4$	

Table 3-6

MINIMUM ROOF CLASSES

OCCUPANCY	TYPES OF CONSTRUCTION						
	I		II		III		IV
	F.R.	F.R.	One-hour	N	One-hour	N	H.T.
A-1	B	B	–	–	–	–	–
A) 2-2.1	B	B	B	–	B	–	B
A-3	B	B	B	B	B	C	B
A-4	B	B	B	B	B	B	B
B	B	B	B	B	B	C	B
E	B	B	B	B	B	B	B
F	B	B	B	B	B	C	B
H-1	A	A	A	A	–	–	–
H) 2-3-4-5-6-7	A	B	B	B	B	B	B
I) 1.1-1.2-2	A	B	B	–	B	–	B
I-3	A	B	B[1]	–	B	–	–
M	B	B	B	B	B	C	B
R-1	B	B	B	B	B	C	B
R-3	B	B	B	B	NR	NR	NR
S-1, S-3	B	B	B	B	B	C	B
S-2, S-5	B	B	B	B	B	B	B
S-4	B	B	B	B	–	–	–
U	B	B	B	B	NR	NR	NR

A – Class A roofing.
B – Class B roofing.
C – Class C roof covering.
NR – Nonrated roof coverings.
N – No requirements for fire resistance.
F.R. – Fire resistive
H.T. – Heavy Timber.

Table 3-7

WOOD SHINGLES	WOOD SHAKES
Not Permitted below 3 Units Vertical in 12 Units Horizontal (25% Slope)	**Not Permitted below 4 Units Vertical in 12 Units Horizontal (33.3% Slope)**
Shingles shall be laid with a side lap of not less than 1 1/2 inches (38 mm) between joints in adjacent courses, and not in direct alignment in alternate courses. Spacing between shingles shall be approximately 1/4 inch (6 mm). each shingle shall be fastened with two nails only, positioned approximately 3/4 inch (19 mm) from each edge and approximately 1 inch (25 mm) above the exposure line. Starter course at the eaves shall be doubled.	Shakes shall be laid with a side lap of not less than 1 1/2 inches (38 mm) between joints in adjacent courses. Spacing between shakes shall not be less than 3/8 inch (9 mm) or more than 5/8 inch (16 mm) except for preservitive-treated wood shakes which shall have a spacing not less than 1/4 inch (6 mm) or more than 3/8 inch (9 mm). Shakes shall be fastened to the sheathing with two nails only, positioned approximately 1 inch (25 mm) from each edge and approximately 2 inches (51 mm) above the exposure line. The starter course at the eaves shall be doubled. The bottom or first layer may be either shakes or shingles. Fifteen-inch or 18 inch (381 mm or 457 mm) shakes may be used for the starter course at the eaves and final course at the ridge.

Table 3-8

ITEM NUMBER	CEILING CONSTRUCTION	THICKNESS OF FLOOR OR ROOF SLAB (inches)				MINIMUM THICKNESS OF CEILING (inches)			
		× 25.4 for mm							
		4 Hr.	3 Hr.	2 Hr.	1 Hr.	4 Hr.	3 Hr.	2 Hr.	1 Hr.
6-1.1	Gypsum plaster on metal lath attached to the bottom cord with single No. 16 gage or doubled No. 18 gage wire ties spaced 6" (153 mm) on center. Plaster mixed 1:2 for scratch coat, 1:3 for brown coat, by weight, gypsum to sand aggregate for two-hour system. For three-hour system plaster is neat.			$2^1/_2$	$2^1/_4$			$^3/_4$	$^5/_8$
6-2.1	Vermiculite gypsum plaster on metal lath attached to the bottom chord with single No. 16 gage or doubled 0.049 inch (1.24 mm) (No. 18 B.W. gage) wire ties 6" (153 mm) on center.		2				$^5/_8$		

Table 3-8 (*Continued*)

ITEM NUMBER	CEILING CONSTRUCTION	THICKNESS OF FLOOR OR ROOF SLAB (inches) × 25.4 for mm				MINIMUM THICKNESS OF CEILING (inches) × 25.4 for mm			
		4 Hr.	3 Hr.	2 Hr.	1 Hr.	4 Hr.	3 Hr.	2 Hr.	1 Hr.
6-3.1	Cement plaster over metal lath attached to the bottom chord of joists with single No. 16 gage or doubled 0.049 inch (1.24 mm) (No. 18 B.W. gage) wire ties spaced 6" (153 mm) on center. Plaster mixed 1:2 for scratch coat, 1:3 for brown coat for one-hour system and 1:1 for scratch coat. $1:1^1/_2$ for brown coat for two-hour system, by weight, cement to sand.				2				$^5/_8$
6-4.1	Ceiling of $^5/_8$" (15.9 mm) Type X wallboard attached to 7/8" deep by $2^5/_8$" (22.2 mm deep by 66.7 mm) by 0.021 inch (0.53 mm) (No. 25 carbon sheet steel gage) hat-shaped furring channels 12"(305 mm) on center with 1" (25.4 mm) long No. 6 wallboard screws at 8" (203 mm) on center. Channels wire tied to bottom chord of joists with doubled 0.049 inch (1.24 mm) (No. 18 B.W. gage) wire or suspended below joists on wire hangers.			$2^1/_2$				$^5/_8$	

Notes

Notes

Chapter

4

Exterior Wall Framing

Loadbearing Exterior Walls

Because the joists that support loadbearing exterior walls are not long enough to run the full width of a house, most framing structures have a central wood girder. The perimeter footing supports the exterior walls as shown in Fig. 4-1. Butted or overlapping floor joists are laid across the girder and one exterior sill. The exterior walls and the walls above the girder then bear the load of the house. See exterior wall cross section in Fig. 4-2. The remaining partition walls are not loadbearing.

Exterior bearing walls including openings (Fig. 4-3) and their anchorage (Fig. 4-4) are designed for a specified force of the effective peak velocity-related acceleration times the weight of the wall, normal to the surface, with a minimum force of 10 percent of the weight of the wall. Interconnection of wall elements and connections to supporting framing systems must have sufficient ductility, rotational capacity, or strength to resist shrinkage, thermal changes, and differential foundation settlement when combined with seismic forces. The architectural symbols for the wall's components are shown in Fig. 4-5.

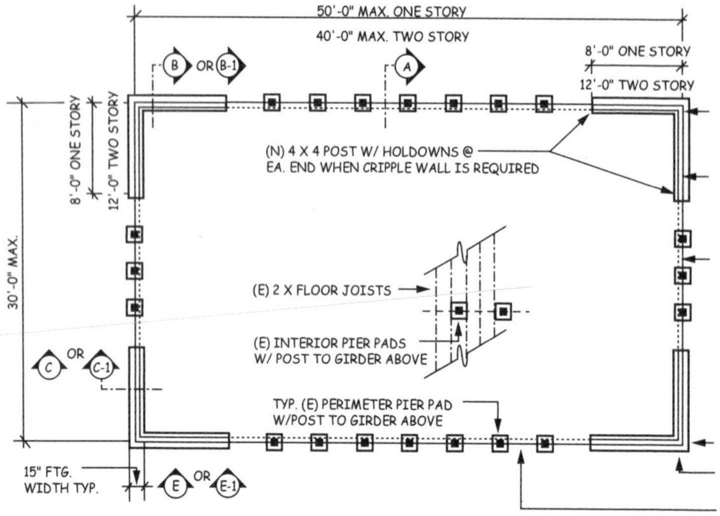

Figure 4-1 Perimeter Footings

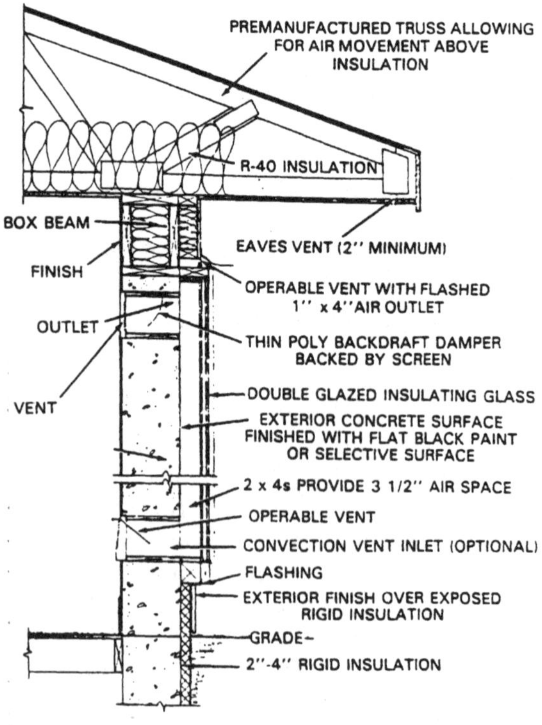

Figure 4-2 Exterior Wall Section

Because shear loads are cumulative as they work their way down an exterior wall, they are greatest at the base of the frame. Shear loads are transferred from the shear wall into the foundation through the sill plate. Consequently, the sill plate must be attached to the foundation with anchor bolts or plates. Before the load enters the anchor bolt or plate, it must successfully pass through the sill plate. See Fig. 4-6. Sill plates on exterior walls should be in good condition and free from cracks. If a cracked or split sill plate is encountered, it should be replaced before proceeding with the wall framing.

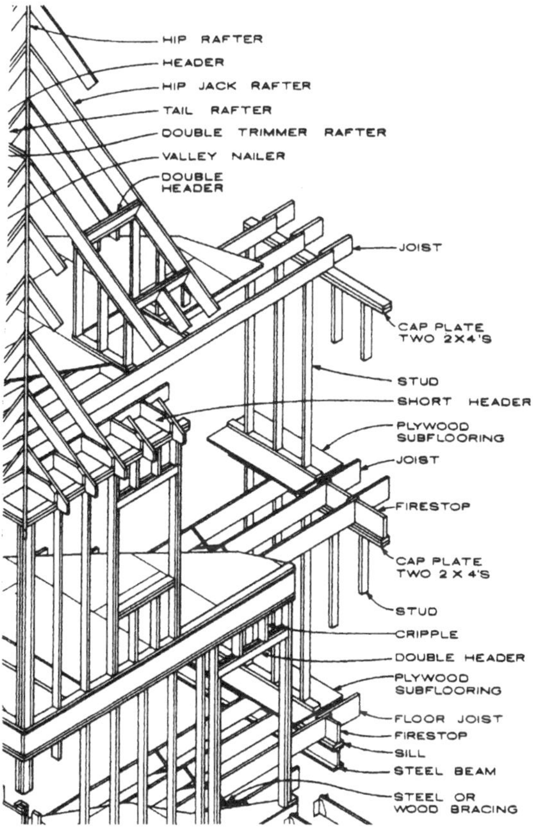

Figure 4-3 Exterior Wall Openings

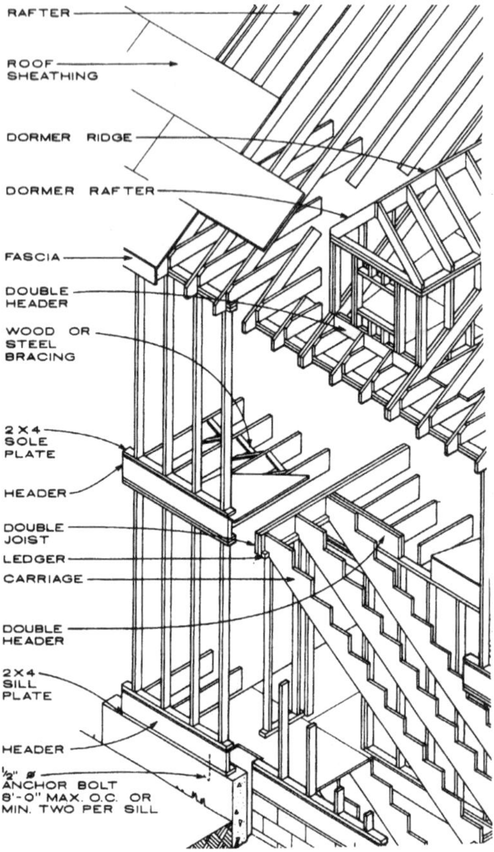

Figure 4-4 Exterior Wall Anchorage

Sometimes in remodeling exterior walls, existing anchor bolts in the sill plate are too few in numbers or size or are rusted through under the sill plate. Some older buildings did not bolt their sill plates to the concrete stem wall. In these cases, special retrofit anchors must be used to strengthen the sill plate connection to the exterior walls. New sill plate anchors are installed in two ways: drilled-in anchors through the sill plate or side plates that connect the sill plate to the foundation concrete or masonry. Side plates are used when there is not enough room to stand a drill motor on top of the sill plate.

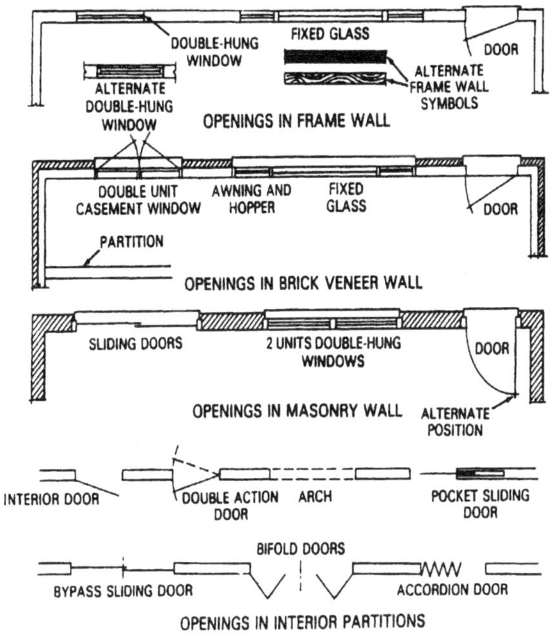

Figure 4-5 Architectural Symbols

Drilled-in sill plate anchors come in two types: mechanical and adhesive. Adhesive anchors have the advantage of working in lower strength existing concrete. They are more expensive to install than mechanical anchors and require greater quality control during installation. Mechanical wedge anchors are easier to install but require generally greater concrete strength due to the concentration of stresses at the expansion clip. When sufficient concrete strength exists, either adhesive or mechanical anchors may be used. See Fig. 4-7. Generally, the strength of both anchor types in concrete is greater than their strength in the wood sill plate.

Supporting columns or piers of inverted pendulum-type structures for exterior bearing walls are designed for the bending moment calculated at the base determined using the procedures given in code tables and vary uniformly to a moment at the top equal to one-half the calculated bending moment at the base.

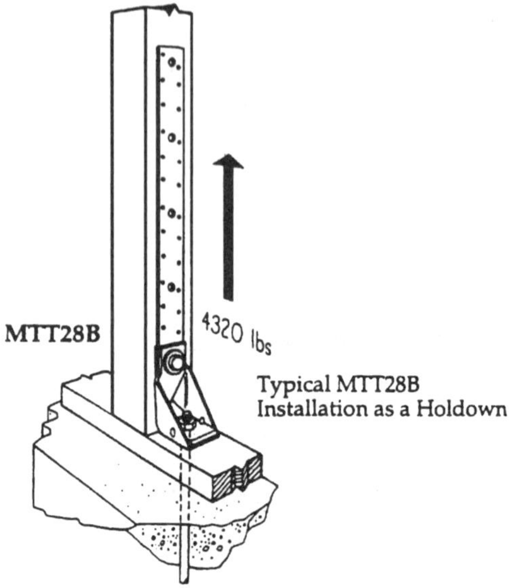

Figure 4-6 Sill Anchor

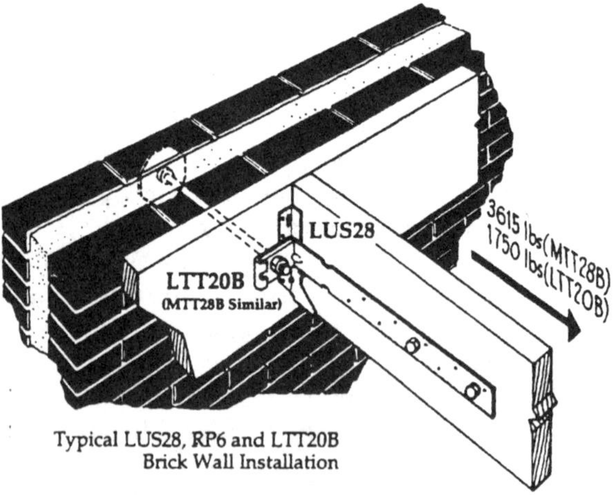

Figure 4-7 Mechanical Anchor

Framing systems that have plan structural irregularity are analyzed for the critical load effect due to direction of application of seismic forces. Alternatively, the building must be analyzed in any two orthogonal directions. Structural elements and foundations are designed for 100 percent of the forces for one direction plus a simultaneous load of 30 percent of the forces for the perpendicular direction.

In these orthogonal load effects, exterior bearing walls are designed for 100 percent of the seismic forces for one direction plus a simultaneous load of 30 percent of the seismic forces for the perpendicular direction. The load combination requiring the maximum structural frame component and assembly strength is used. Diaphragms and components of the seismic resisting system utilized in only one of the two orthogonal directions are not required to be designed for the combined load effects.

In designing loadbearing framing that will resist vertical seismic loads, the vertical component of earthquake ground motion is considered in the design of horizontal cantilever and horizontal prestressed components. Horizontal prestressed components are designed for load combinations in accordance with local adopted building code specs and horizontal cantilever structural components are to be designed for a net upward force of 0.2 times the dead load in addition to the applicable load combinations.

For frames with significant torsional deflections, the maximum drift includes torsional effects. The total deflection of an exterior bearing wall due to seismic design forces cannot encroach on an interior nonbearing wall. All portions of the framing system must be designed and constructed to act as an integral unit in resisting seismic force unless separated structurally by a distance sufficient to avoid contact that would damage the structural system of the building under total deflection. See Table 4-1.

Bearing Walls

In bearing walls, the wall studs must be capped with a double top plate installed to provide overlapping at corners and

Table 4-1

DOUG FIR-LARCH ALLOWABLE LOADS					SPRUCE-PINE-FIR ALLOWABLE LOADS				
UPLIFT	FLOOR (100)	ROOF			UPLIFT	FLOOR (100)	ROOF		
		SNOW (115)	CONST (125)	WIND (133)			SNOW (115)	CONST (125)	WIND (133)
470	640	735	800	855	390	545	625	680	725
935	835	960	1045	1115	780	705	810	880	935
1550	2565	2950	3205	3315	1390	2155	2475	2540	2580
935	1015	1170	1270	1355	785	860	990	1075	1150
2000	3315	3430	3505	3570	1855	2580	2680	2745	2800
935	1280	1470	1600	1705	780	1085	1250	1360	1450
2845	3505	3650	3750	3830	2320	2745	2870	2955	3025
565	765	880	960	1020	465	640	735	800	855
1140	1000	1150	1250	1335	925	820	945	1025	1095
1080	1005	1155	1255	1340	875	825	950	1030	1100
1550	2580	2965	3255	3440	1395	2165	2490	2710	2890
2255	3695	4250	4620	4835	1840	3125	3590	3905	4165
1075	1265	1455	1585	1690	875	1050	1210	1315	1405
1550	1505	1730	1885	2010	1315	1240	1425	1550	1650
2000	3885	4465	4855	5180	1860	3275	3765	4095	4370
2650	6140	7060	7675	8000	2300	5225	6010	6125	6195
1550	1765	2030	2210	2355	1390	1465	1680	1830	1950
2160	2010	2310	2510	2680	1755	1650	1900	2065	2200
2855	5190	5900	5900	5900	2330	4385	4795	4875	4945
3385	7750	8085	8195	8290	2760	6085	6235	6330	6415
1550	2030	2335	2540	2710	1390	1695	1945	2115	2255
540	765	880	960	1020	435	640	735	800	855
1075	1000	1150	1250	1335	875	820	945	1025	1095
1080	1005	1155	1255	1340	875	825	950	1030	1100
1550	2580	2965	3225	3440	1395	2165	2490	2710	2890
1075	1265	1455	1585	1690	875	1050	1210	1315	1405
1550	1505	1730	1885	2010	1315	1240	1425	1550	1650
2000	3885	4465	4855	5180	1860	3275	3765	4095	4370
2650	6140	7060	7675	8000	2300	5225	6010	6530	6965
1550	1765	2030	2210	2355	1310	1465	1680	1830	1950
2160	2010	2310	2510	2680	1755	1650	1900	2065	2200
2855	5190	5900	5900	5900	2300	4385	5040	5480	5585
3385	7750	8665	8665	8665	2760	6605	7085	7180	7265
3945	9230	9430	9560	9670	3220	7100	7270	7385	7485
1550	2030	2335	2540	2710	1390	1695	1945	2115	2255
4510	9420	9645	9800	9925	3680	7265	7460	7590	7700

Notes: All values are in pounds per square foot (lb/ft^2).

intersections with bearing partition walls. End joints in top plates are typically offset at least 48 in (1219 mm). The exception here is that a single top plate may be installed in bearing and exterior walls provided the plate is adequately tied at all joints, corners, and intersecting walls by a 3 × 6 × 0.036 in-thick (76 × 153 × 0.914 mm) galvanized steel plate that is nailed to each wall or segment of wall by three 8d nails, provided the rafters or joists are centered over the studs with a tolerance of no more than 1 in (25 mm). The top plate may be omitted over lintels that are adequately tied to adjacent wall sections with steel plates or equivalent as previously described (check local code specs). See Fig. 4-8.

Placement of the framing members connecting the load path to bearing studs is crucial to the design strength of exterior walls. Where floor or roof framing members are spaced more than 16 in (406 mm) on center and the bearing studs are spaced 24 in (610 mm) on center, the framing members must bear within 5 in (127 mm) of the studs. The three exceptions to this general rule of thumb are: if the top plates are two 2 × 6 × 0.036 in-thick (76 × 153 × 0.914 mm), solid blocking equal in size to the studs is installed to reinforce the double top plate; or a third top plate is installed.

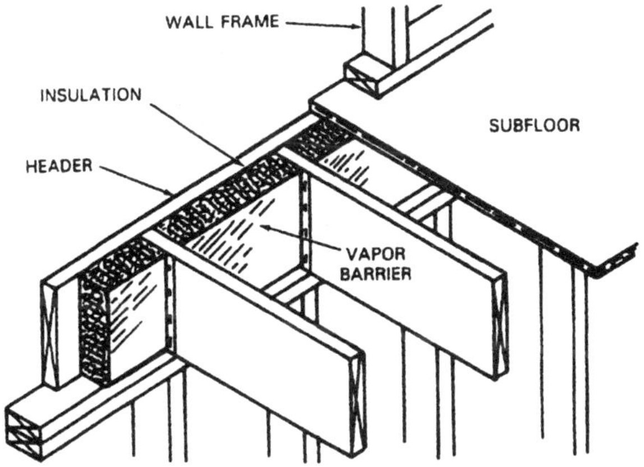

Figure 4-8 Top Plates Supporting Second Story

Interior loadbearing partition walls are constructed, framed, and firestopped as specified for exterior walls. Interior nonbearing partition walls may be constructed with 2 × 3 in (51 × 76 mm) studs spaced 24 in (610 mm) on center or 2 × 4 in (51 × 102 mm) flat studs spaced 16 in (406 mm) on center as shown in Table 4-2. Interior nonbearing partitions may be capped with a single top plate.

Any stud in an exterior wall or bearing interior partition may be cut or notched to a depth not exceeding 25 percent of the total stud width. Studs in nonbearing partitions may be notched to a depth not exceeding 40 percent of a single stud width. Any stud may be bored or drilled, provided that the diameter of the resulting hole is no greater than 40 percent of the stud width, the edge of the hole is no closer than $^5/_8$ in (15.9 mm) to the edge of the stud, and the hole is not located in the same section as a cut or notch. Wall-bracing specifications are shown in Table 4-3.

Framing for Increased Thermal Insulation

The industry standard 2 × 4 (38 × 89 mm) has been the standard wall stud since light framing was invented. In recent years, however, pressures for heating fuel conservation have created a need to provide more thermal insulation than can be inserted in the cavities of a wall framed with 2 × 4s. Some building designers have simply adopted the 2 × 6 (38 × 140 mm) as the standard stud, usually at an on center spacing of 24 in (610 mm). Others have stayed with the 2 × 4 stud but have covered the wall either inside or out with insulating plastic foam sheathing, thus reaching an insulating value about the same as that of a 2 × 6 wall.

Many use both 2 × 6 studs and insulating sheathing. Building designers in some very cold climates are not satisfied even with these types of construction and use exterior walls framed in two separate layers, or with vertical truss studs made up of two ordinary studs joined at intervals by plywood plates.

Table 4-2
Fastening Schedule

Fastening Schedule

Connection	Fastener	Number or Spacing
2-inch planks	16d common	2 each bearing
Studs to sole plate, end nail	16d common	2 each end
Wood Structural Panel and Particleboard Subflooring		
15/32", 1/2"	6d common, annular or spiral thread	6" o.c. edges and 12" o.c. intermediate
19/32"-3/4"	8d common or 6d annular or spiral thread	6" o.c. edges and 12" o.c. intermediate
1", 1 1/8"	10d common or 8d annular or spiral thread	6" o.c. edges and 6" o.c. intermediate
15/32", 1/2"	16 ga galvanized wire staples, 3/8" minimum crown 1 5/8" length	4" o.c. edges and 7" o.c. intermediate
19/32", 5/8"	16 ga galvanized wire staples, 3/8" minimum crown 1 5/8" length	2 1/2" o.c. edges and 4" o.c. intermediate
Wood Structural Panel and Particleboard Roof & Wall Sheathing		
1/2" or less	6d common	6" o.c. edges and 12" o.c. intermediate
19/32" or greater	8d common	6" o.c. edges and 12" o.c. intermediate
5/16" - 1/2"	16 ga galvanized wire staples, 3/8" min. crown. Length of 1" plus wood structural panel or particleboard thickness	4" o.c. edges and 8" o.c. intermediate

Table 4-3

SOUTHERN PINE ALLOWABLE LOADS					SPRUCE-PINE-FIR ALLOWABLE LOADS							
LENGTH OF BOLT IN CARRYING MEMBER	FLOOR (100)	ROOF			UPLIFT[1,2]			LENGTH OF BOLT IN CARRYING MEMBER	FLOOR (100)	ROOF		
		SNOW (115)	CONST (125)	WIND (133)	MIN NAILS	MAX NAILS	BOLTS			SNOW (115)	CONST (125)	WIND (133)
1½	1395	1605	1745	1860				1½	1180	1360	1475	1575
3	2745	3155	3430	3660	555	1940	2345	3	2310	2660	2890	3080
4½	2965	3410	3705	3955				4½	2735	3145	3420	3650
6	2955	3400	3695	3940				6	2730	3140	3410	3635
1½	2030	2335	2540	2705				1½	1725	1985	2155	2300
3	4120	4740	5150	5495	590	2370	3515	3	3475	4000	4345	4635
4½	4415	5080	5520	5890				4½	4080	4695	5105	5445
6	4395	5055	5495	5860				6	4060	4670	5075	5415
1½	—	—	—	—				1½	—	—	—	—
3	7120	8190	8900	9495	590	2665	6475	3	6005	6700	6700	6700
4½	10000	10000	10000	10000				4½	6700	6700	6700	6700
6	10000	10000	10000	10000				6	6700	6700	6700	6700

Notes: All values are in pounds per square foot (lb/ft^2).

Exterior Wall Finishing

As a platform frame nears completion, exterior finishing operations begin (Figs. 4-9 to 4-11). First the roof edges are finished; then the roof is shingled to offer as much weather protection as possible to subsequent operations.

When the roof is tight to the weather, the windows, doors, and siding are applied, after which interior finishing work may take place completely sheltered from sun, rain, snow, and wind. The outside of the building is now complete for all practical purposes; exterior painting, finish grading, landscaping, and paving work may commence.

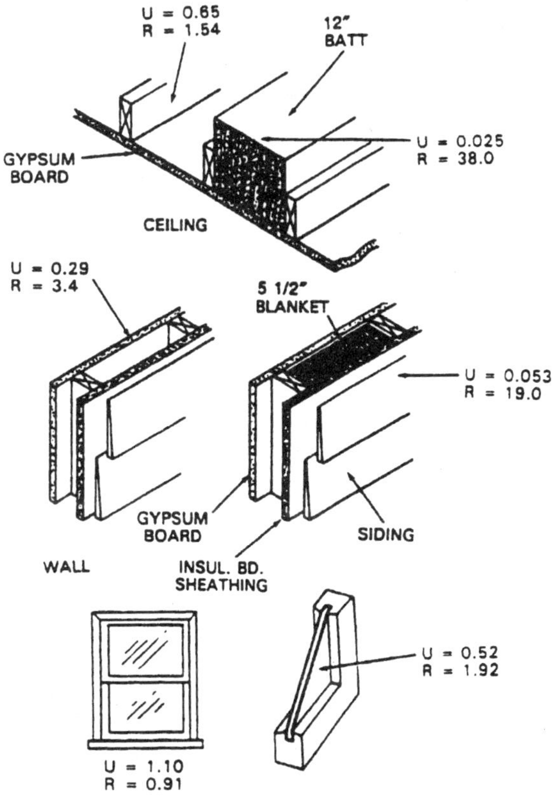

Figure 4-9 Typical Insulating Values

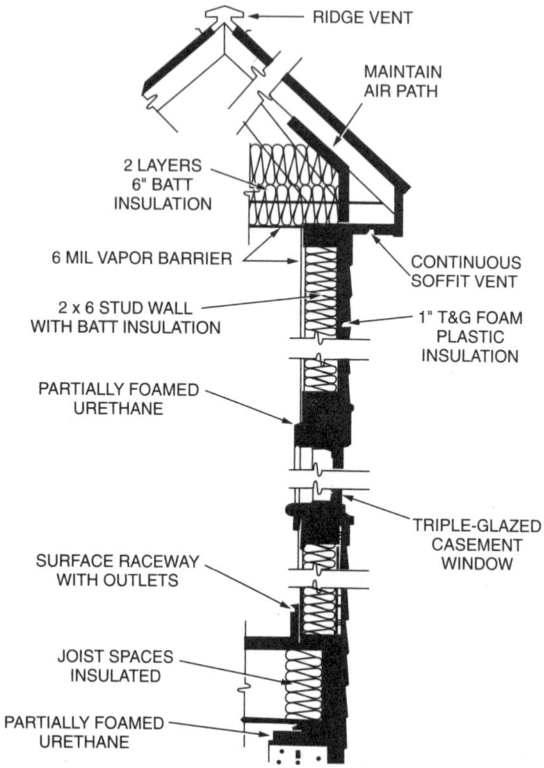

Figure 4-10 Exterior Finishing

The allowable spans and loads for exterior walls with wood structural panel sheathing are shown in Table 4-4. Plywood exterior wall coverings are regulated by the thicknesses shown in Table 4-5. Particleboard wall sheathing and subfloor specifications are shown in Table 4-6.

Windows

Windows were formerly made on the job site by highly skilled carpenters but are now produced almost exclusively in factories. Some manufacturers make a range of standard sizes from which the designer can select, while others build windows to order.

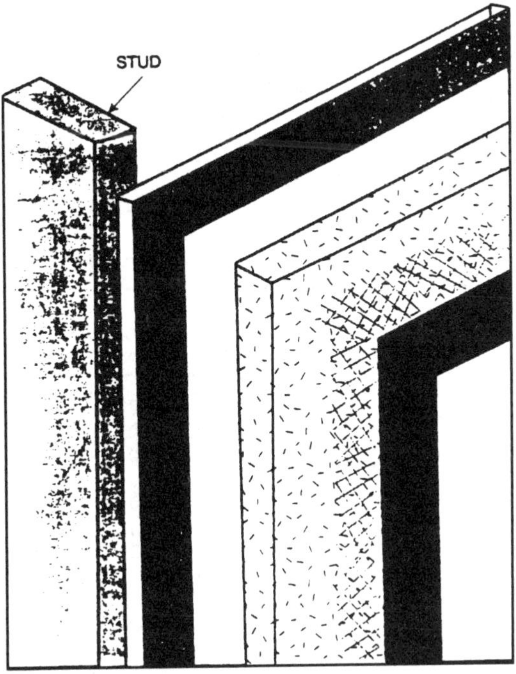

Figure 4-11 Exterior Finishing

The rationale for factory production in either case is one of higher efficiency, lower cost, and most importantly, better quality. Windows need to be made to a very high standard of precision if they are to operate easily and maintain a high degree of weathertightness over a period of many years.

In cold climates especially, a loosely fitted window with single glass and a frame that is highly conductive of heat will significantly increase heating fuel consumption for a building, cause noticeable discomfort to the people in the building, and create large quantities of condensate to stain and decay finish materials in and around the window.

Types of Framed Windows

Fixed windows are the least expensive and the least likely to leak air or water because they have no operable components.

Table 4-4

Allowable Spans and Loads for Wood Structural Panel Sheathing and Single Floor Grades Continuous Over Two or More Spans With Long Dimension Perpendicular to Supports[1,6]

Sheathing Grade		Roof				Floor
Span Rating	Panel Thickness (in)	Maximum Span (in)		Load (psf)		
Roof/Floor Span		With Edge Support[2]	Without Edge Support	Total Load	Live Load	Maximum Span (in)
12/0	5/16	12	12	40	30	0
16/0	5/16, 3/8	16	16	40	30	0
20/0	5/16, 3/8	20	20	40	30	0
24/0	3/8, 7/16, 1/2	24	20[3]	40	30	0
24/16	7/16, 1/2	24	24	50	40	16[5]
32/16	15/32, 1/2, 5/8	32	28	40	30	16[4,5]
40/20	19/32, 5/8, 3/4, 7/8	40	32	40	30	20
48/24	23/32, 3/4, 7/8	48	36	45	35	24
54/32	7/8, 1	54	40	45	35	32
60/48	7/8, 1, 1-1/8	60	48	45	35	48
Single Floor Grade		**Roof**				**Floor**
		Maximum Span (in)		**Load (psf)**		
Span Rating	**Panel Thickness (in)**	**With Edge Support[2]**	**Without Edge Support**	**Total Load**	**Live Load**	**Maximum Span (in)**
16 o.c.	19/32, 5/8	24	24	50	40	16[5]
20 o.c.	19/32, 5/8, 3/4	32	32	40	30	20[4,5]
24 o.c.	23/32, 3/4	48	36	35	25	24
32 o.c.	7/8, 1	48	40	50	40	32
48 o.c.	1-3/32, 1-1/8	60	48	50	40	48

1 in = 25.4 mm
1 psf = 47.8803 Pa

Notes:

1. The allowable loads were determined using a dead load of 10 psf. If the dead load exceeds 10 psf then the live load shall be reduced accordingly.
2. Tongue-and-groove edges, panel edge clips (one midway between each support, except two equally spaced between supports 48 inches on center), lumber blocking, or other. Only lumber blocking shall satisfy blocked diaphragm requirements.
3. For 1/2-inch panels maximum span shall be twenty-four inches.
4. Maximum framing space shall be 24 inches on center for floors where 1-1/2 inches of cellular or lightweight concrete is applied over the panels.
5. Maximum frame spacing shall be 24 inches on center where 3/4-inch wood strip flooring is installed at right angles to joist.
6. Shall apply only to panels 24 inches or wider.

Table 4-5

Allowable Spans for Wood Structural Panel Wall Sheathing			
		Maximum Stud Spacing (in) with Exterior covering nailed to:	
Panel Span Rating	Panel thickness (in) and Construction	Stud	Sheathing
12/0, 16/0, 20/0 or Wall-16oc	5/16	16	16
24/0, 24/16, 32/16 or Wall-24oc	3/8, 7/16, 15/32, 1/2	24	24

1 in = 25.4 mm

Table 4-6

Minimum Net Thickness of Lumber Subflooring (in)		
Joist Spacing (in)[1]	Perpendicular To Joists	Diagonally To Joists
24	11/16	3/4
16	5/8	5/8
12	5/8	5/8

1 in = 25.4 mm

Note:

1. Joints in subflooring shall occur over supports unless end-matched lumber is used, in which case each piece shall bear on at least two joists.

Single-hung and *double-hung* windows have one or two moving *sashes*, which are the frames in which the glass is mounted. The sashes slide up and down in tracks in the frame of the window. In older windows, the sashes were held in position by cords and counterweights, but today's double-hung windows rely on a system of springs to counterbalance the weight of the sashes.

A *sliding window* is essentially a single-hung window on its side, and shares with single-hung and double-hung windows the advantage that the sashes are always securely held in tracks in the frame. This inherently stable construction allows single-hung, double-hung, and sliding windows to be designed in an almost unlimited range of sizes and proportions. It also allows the sashes to be more lightly built than those in *projected windows*, which have sashes that pivot in their frames and must therefore have enough structural stiffness to resist wind loads. Windows with sashes that slide are also easier to make tight against the penetration of wind and water, but none can be opened to more than half its total area.

All projected windows can be opened to their full area. *Casement windows* are helpful in catching passing breezes and inducing ventilation through the building; they are generally narrow in width but can be joined to one another and to sashes of fixed glass to fill wider openings. *Awning windows* can be broad but are not usually very tall. Awning windows have the advantages of protecting an open window from water during a rainstorm and of lending themselves to a building-block approach to the design of window walls. *Hopper windows* are rare in residential construction but are more common in commercial work.

Windows in roofs are specially constructed and flashed for watertightness and may be either fixed (*skylights*) or openable (*roof windows*). Large glass doors may slide in tracks or swing open on hinges. The hinged *French door* opens fully and is a more generous type of door than the sliding door, but it cannot be used to regulate airflow through the room unless it is fitted with a plunger-type doorstop that can hold it open in any position. The French door, with its seven separate edges that must be carefully fitted and weather-stripped, is prone to air leakage along these edges. The *terrace door* is a recently developed door that minimizes this problem, but, like the sliding door, it can open to only half its area.

Window Frames

The traditional frame material for windows, and in many ways still the best, is wood—though aluminum, steel, plastics, and combinations of these four materials have also come into widespread use. Wood is a fairly good thermal insulator and is easily worked into sash, but it shrinks and swells with changing moisture content and requires repainting every few years.

Aluminum is inexpensive and easy to work as a sash material and requires no periodic repainting. But unless *thermally broken* with plastic or synthetic rubber components to interrupt the flow of heat through the metal, aluminum conducts heat so well that condensate and sometimes even frost will form on interior window frame surfaces during cold winter weather.

The chief advantage of steel is its strength—steel sash sections are slender and unobtrusive compared to those of wood and aluminum. Unless permanently coated, steel windows need repainting at intervals. Plastic window frames never need painting and conduct little heat. But they must be designed carefully to accommodate the large thermal expansion and contraction movements that are characteristic of plastics.

A number of manufacturers produce windows whose frames consist of a core of wood covered with an exterior layer of either plastic or aluminum. These are designed to combine the strength, stiffness, and low thermal conductivity of wood with the long-lasting, maintenance-free characteristics of aluminum and plastic. The wood is often left exposed on the interior of the frame so it may be painted or stained in the conventional way.

The structural wood framing for window openings is shown in Figs. 4-11 and 4-12.

Exterior Wall Glazing and Screening

Factory-made windows are extremely easy to install, often requiring only a few minutes per window. Interior trimming of a window takes longer. Single glazing on exterior walls is acceptable only in the mildest climates because of its low resistance to heat flow. Storm windows or double glazing have become the minimum acceptable glazing under most building codes. The storm window has a slight thermal

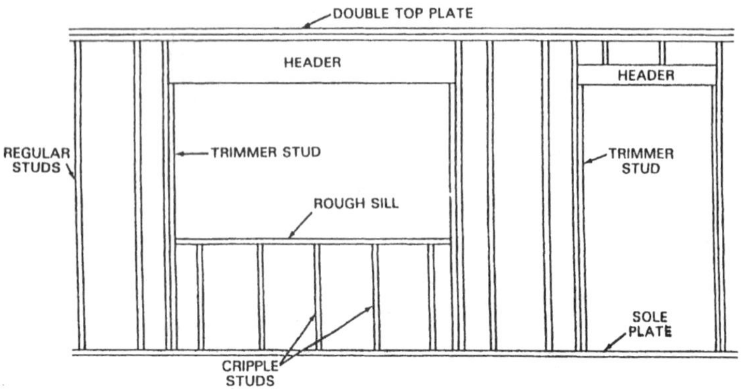

Figure 4-12 Openings Framing

advantage over double glazing mounted in the movable sash because it acts as a second seal against air leakage around the edges of the sash, which is typically a major source of heat loss. See Fig. 4-13.

Removable glazing panels and storm windows must be removed periodically and cleaned, which is a nuisance that can be avoided by using sealed double or triple glazing. Sealed triple glazing is often heavy enough to cause problems with the operation of the sash but performs well from a thermal standpoint. Some manufacturers avoid the weight problem by using a thin, virtually weightless, stretched plastic film for the middle glazing sheet.

Double glazing with an internal coating that is selectively reflective of longer-wave infrared radiation performs at least as well as triple glazing, though it still transmits heat five to ten times as rapidly as an equal area of well-insulated wall. This *low emissivity coating* allows nearly all wavelengths of sunlight to enter the building but blocks radiation of heat from the interior of the building to the outdoors.

The thermal performance of a window in a cold climate can be further improved by filling the space between the glazing sheets with an inert gas such as argon that is a poor transmitter of heat. In warm climates, windows are often provided with tinted glass that reduces the transmission of

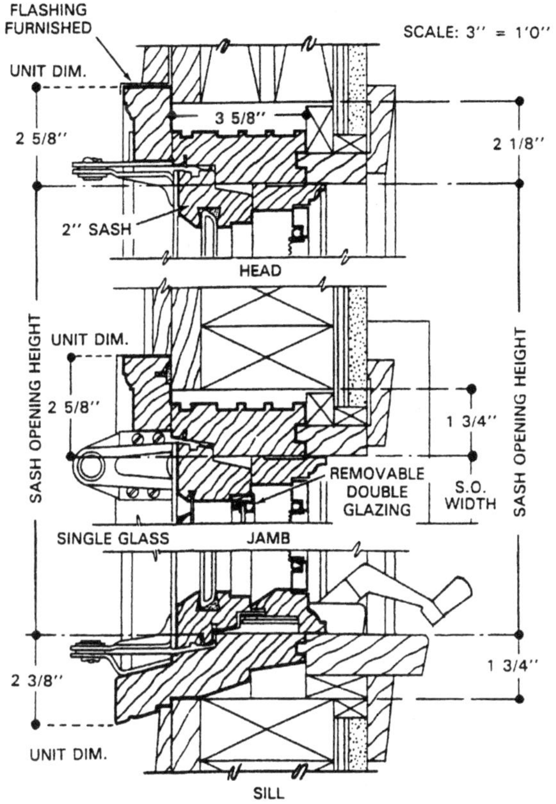

Figure 4-13 Sash Cross Section

solar heat and makes the building easier to cool.

Large windows that are near enough to the floor that they might be mistaken for open doorways must be glazed with breakage-resistant material under most building codes. Tempered glass is most often used for this purpose, but laminated glass and plastic glazing sheets are also permitted.

Insect screen arrangements vary from one type of window to another. Single-hung, double-hung, and sliding windows generally have removable or retractable screens on the outside. Projected windows must have interior screens. Sliding patio doors and terrace doors have an exterior sliding screen, while French doors require a pair of hinged screen doors on the exterior.

Doors

Exterior doors must be well constructed and tightly weather-stripped if they are not to leak air and water. The problem can be compounded by the seasonal expansion and contraction of wood doors. Properly installed and finished wood panel or solid-core doors are excellent for exterior use. Wood exterior doors exposed to direct sunlight should not be painted or stained in dark colors, which absorb so much solar heat that the wood in the door dries unevenly and warps. Casing for interior wood doors is shown in Fig. 4-14. Plumbing the door jambs is shown in Fig. 4-15.

Pressed sheet-metal doors are becoming popular as an alternative to wood exterior doors. Their hollow cores are filled with insulating plastic foam.

Foam-filled metal doors are superior thermally to wood doors. They do not suffer from moisture expansion and contraction and are often furnished *prehung*, meaning that they are already mounted on hinges in a frame, complete with weather-stripping, ready to install by merely nailing the frame in place.

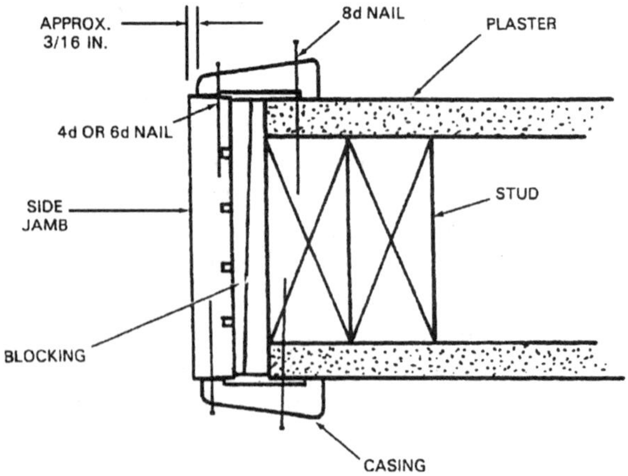

Figure 4-14 Door Casing

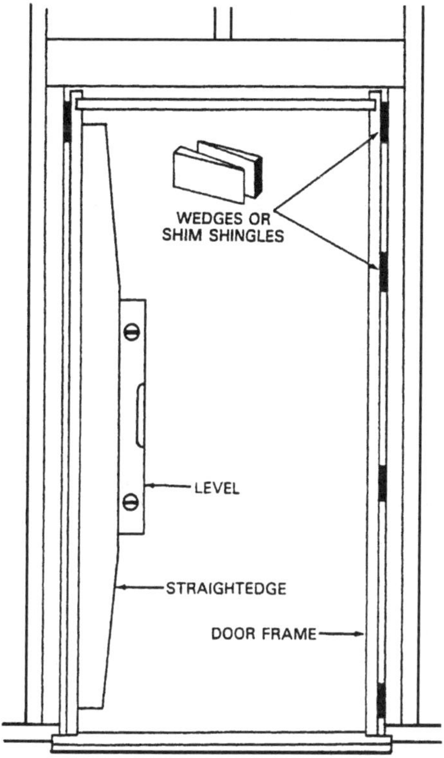

Figure 4-15 Plumbing Doors

Wood doors can also be purchased prehung, although many are still hung and weather-stripped on the building site. The major disadvantage of metal exterior doors is that they do not have the satisfying appearance, feel, or sound of a wood door.

Glass in or near doors is *tempered* in order to strengthen it to meet building code safety requirements. Door manufacturers furnish tempered glass in doors as a matter of course, but the designer who specifies locally custom-fabricated doors may have to pay particular attention to see that this requirement is met.

Residential entrance doors almost always swing inward and are mounted on the interior side of the frame. For an improved wintertime thermal performance of the entrance, a *storm door* is often mounted on the outside of the frame, swinging outward. The storm door usually includes at least one large panel of tempered glass. In summer, a *screen door* may be substituted for the storm door.

A *combination door*, which has easily interchangeable screen and storm panels, is more convenient than separate screen and storm doors.

Exterior Wall Siding

The exterior cladding material applied to the walls of a wood light frame building is referred to as *siding*. Many different types of materials are used as siding: wood boards with various profiles, applied either horizontally or vertically; plywood; wood shingles; metal or plastic sidings; brick or stone; and stucco.

Siding is applied over a paper or felt layer, which acts as an air barrier and backup waterproofing layer. This layer should allow water vapor to pass freely so it does not accumulate in the wall. The traditional material for this layer is asphalt-saturated felt, but airtight, vapor-permeable papers made of synthetic fibers are increasingly used. Construction technology in the area of composite materials exterior siding is changing rapidly at the time of this writing.

Board Siding

Horizontally applied board sidings, made either of solid wood or of wood composition board, are usually nailed into the studs. They can be applied directly over insulating sheathing materials without requiring a *nail-base sheathing*, a material such as plywood or waferboard that accepts and holds nails.

Wood exterior wall siding patterns known as rustic drop siding or shiplap must have an average thickness of not less than $^{19}/_{32}$ in (15.1 mm) and a minimum thickness of not less than $^{3}/_{8}$ in (9.5 mm). Bevel siding must have a minimum thickness measured at the butt section of not less than $^{7}/_{16}$

in (11.1 mm) and a tip thickness of not less than $^3/_{16}$ in (4.8 mm). Siding of lesser dimensions may be used provided such wall covering is placed over sheathing that conforms to the provisions of the local code.

Board siding applied vertically, as seen in Fig. 4-16, must be nailed to horizontal nailing strips or blocking set 24 in (610 mm) on center. The nails must penetrate $1^1/_2$ in (38 mm) into studs, blocking, studs or blocking and sheathing combined, or nailing strips. Horizontal siding, as seen in Fig. 4-17, typically falls on 8-ft centers. The siding application is shown in Fig. 4-18.

Exterior wall structural panels must be of the exterior type and have a thickness of $^3/_8$ in (9.5 mm), except as may be provided by local code. All wood structural panel joints must be backed solidly with nailing pieces not less than 2 in (51 mm) wide, unless wood, wood structural panel, or particleboard sheathing is used or otherwise made waterproof as required by local code. Exterior wall structural panels will align according to the foundation plan, as shown in Fig. 4-19.

The mounting framework is not typically required to be protected when the joints are protected by a continuous wood batt, caulking, flashing, or vertical or horizontal shiplap. Fiberboard siding must be of medium density not less than $^1/_2$ in (12.7 mm) nominal thickness.

Particleboard siding used for covering the exterior of outside walls must be of the Exterior Type 2-M grades conforming to ANSI A208.1. Nails must be spaced not less than $^3/_8$ in (9.5 mm) from edges and ends. Joints must occur over framing members unless particleboard panel siding is applied over $^5/_8$ in (15.9 mm) net wood sheathing or $^{15}/_{32}$ in (11.9 mm) plywood or $^1/_2$ in (12.7 mm) particleboard sheathing. The framework must be protected as required by local code unless the joints are protected with a continuous wood batt, caulking, flashing, or vertical or horizontal shiplap.

Exterior wall wood siding of not less than 1 in (25.4 mm) nominal thickness, $^7/_{16}$-in (11.1-mm) exterior hardboard siding or $^3/_8$-in (9.5-mm) exterior type wood structural panels or particleboard may be used on exterior walls when all the following conditions are met:

Figure 4-16 Vertical Board Siding

Figure 4-17 Horizontal Board Siding

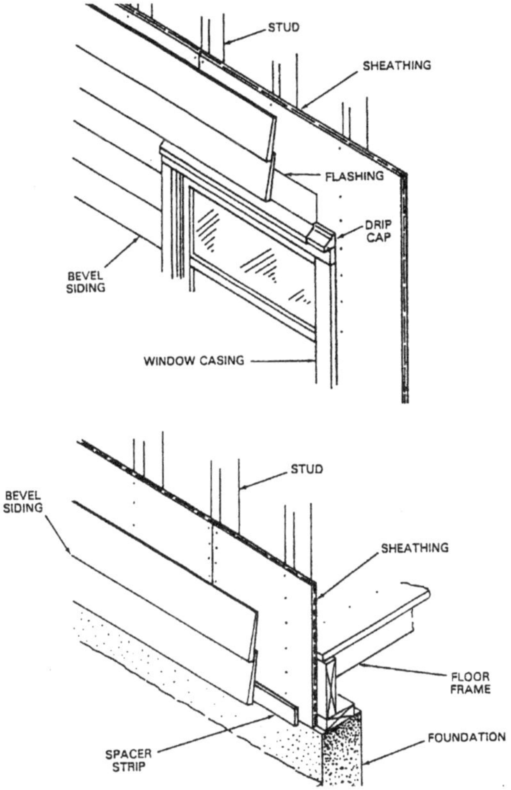

Figure 4-18 Siding Application

- The wall to which the siding is attached faces a street or a permanent open space of 30 ft (9144 mm) or more wide
- The siding does not exceed two stories in height, measured from grade, except when it is fire-retardant-treated for exterior use
- The siding is attached to or furred from a noncombustible backing of the fire resistance required
- Where open or spaced wood siding (without concealed spaces) is used, it cannot project more than 24 in (610 mm) from the framing wall

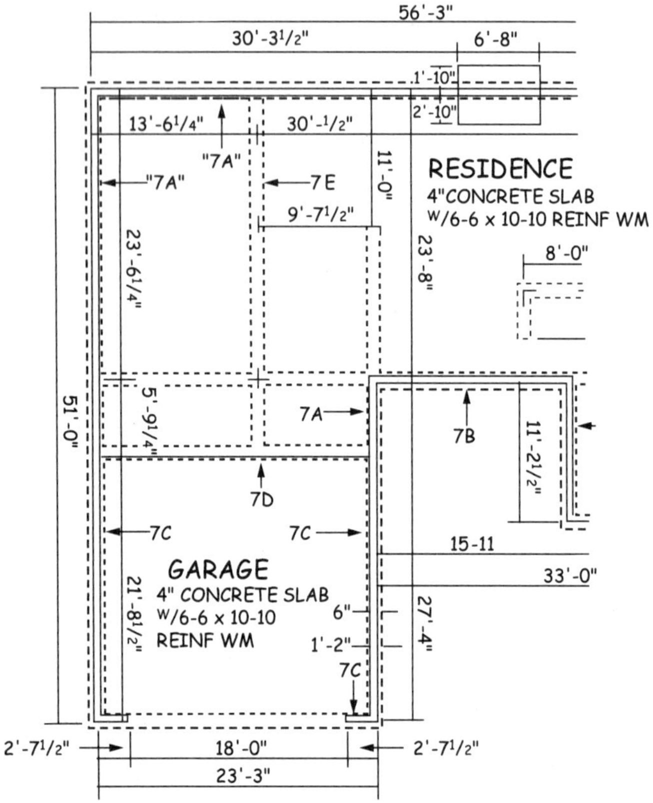

Figure 4-19 Foundation Plan

Where the wood siding is furred from the wall and forms a solid surface, the distance between the back of the siding and the wall cannot exceed $1^5/_8$ in (41 mm) and the space thereby created must be fireblocked during framing and arranged so that there will be no open space exceeding 100 sq ft (9.3 m^2).

Where wood furring strips are used, they must be of approved wood of natural decay resistance or pressure-treated wood.

Asbestos shingles attached to exterior wall sheathing other than wood, plywood, or 2-M-W particleboard must be secured with approved mechanically bonding nails or by corrosion-resistant common nails on shingle nailing boards securely nailed to each stud with two 8d nails, except that asbestos shingles may be attached directly to fiberboard nail base sheathing with corrosion-resistant annular grooved nails. Asbestos shingles must have a minimum thickness of $^5/_{32}$ in (4 mm).

Exterior Wall Veneer Surfacing

The major model national building codes incorporate recognized tested and applied industry standards for use in judging the performance of exterior walls veneer surfacing, the standard materials, systems, and installations. This provides for a system that uses the equal treatment of both proven traditional materials and systems, and the introduction of new construction technologies.

The building codes, in this manner, provide for the efficient introduction of new materials into the traditional construction process. The purpose here is to assure a high level of public safety and construction technologies advancement.

This section covers the construction of exterior walls, architectural trim, balconies, bay windows, veneer, and openings for fire department access. Veneer refers to a facing of the framing with brick; tile; concrete; masonry units; metal, including metal coated with porcelain enamel; glass; wood; or similar material securely attached to a wall for the purpose of providing ornamentation, protection, or insulation but not so bonded as to exert a common reaction under load.

Figure 4-20 shows a cross-sectional view of a typical exterior brick veneered wall. Veneer is not assumed to support any load other than its own weight, nor is it assumed to add to the strength of the wall. Brick veneer construction is shown in Fig. 4-21.

Flashing must be provided as necessary to prevent the entrance of water at openings in or projections through exte-

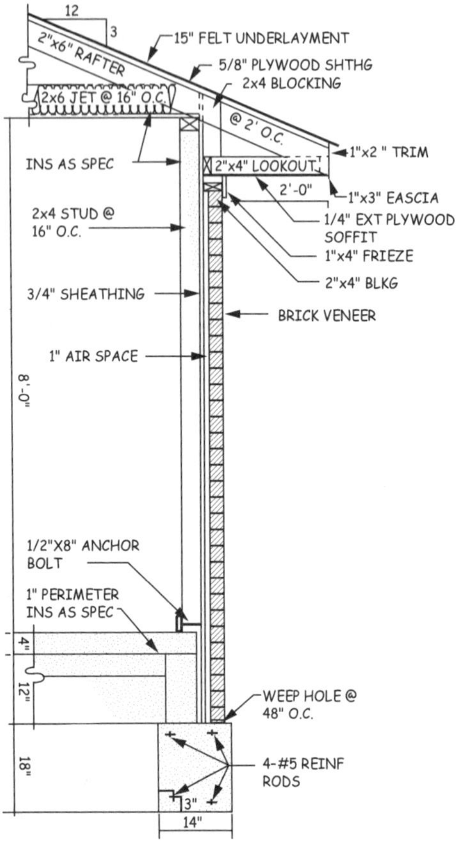

Figure 4-20 Exterior Wall Veneer

rior walls. Flashing is provided at intersections of surfaced exterior walls of different materials unless such materials provide a self-flashing joint and at other points subject to the entrance of water. Caulking must be provided where such flashing is determined by the building official to be impractical.

Exterior anchored veneer supported laterally by the wood frame is limited to a maximum height of 30 ft (9144 mm). Anchored veneer installed more than 30 ft (9144 mm) in height above the noncombustible foundation or support

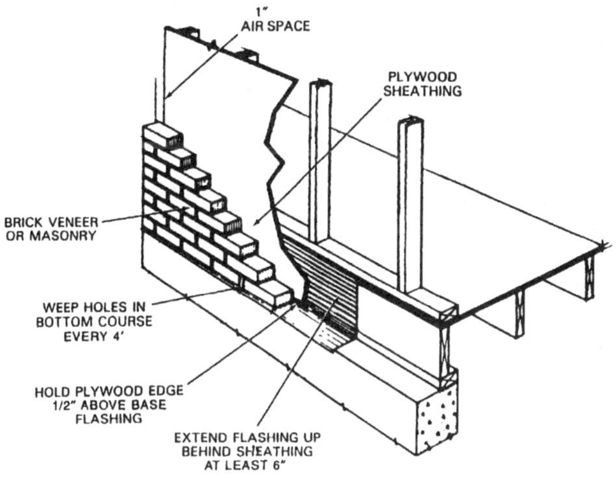

Figure 4-21 Brick Veneer

must be laterally supported by noncombustible structural framing.

The structural framing must horizontally support the weight of veneer at least at the initial 30-ft (9144-mm) height and at each story height thereafter. These height restrictions may be increased for noncombustible structural framing when special design techniques, approved by the local building official, are used in construction of the frame's exterior walls. Noncombustible lintels and noncombustible supports are provided over all openings where the anchored veneer is not self-supporting.

Any masonry veneer anchored to exterior wall framing must be attached with corrosion-resistant corrugated sheet metal that is not less than 0.029 in (0.74 mm) by $^7/_8$ in (22.2 mm) wide or corrosion-resistant ties of strand wire that are not less than No. 9 gauge (3.76 mm) wire with ends of the wire bent to a 90° (1.6 rad) angle to form a hook not less than 2 in (51 mm) long. The metal ties must be embedded in the mortar joint a minimum of one-half the veneer thickness and each metal tie must support not more than 3 sq ft (0.28 m^2) of wall area with a maximum spacing of 16 in (406 mm) vertically and 32 in (813 mm) horizontally.

When anchored veneer is applied over the wood frame, the exterior wall studs must be spaced a maximum of 24 in (610 mm) on center horizontally and be faced with sheathing materials or insulation board on both sides. The minimum thickness of the sheathing material or insulation board installed on each side of the studs must be $^1/_2$ in (12.7 mm). In addition, a 1-in (25.4-mm) minimum air space must be maintained between the anchored veneer and the exterior face of the sheathing material or insulation board. Moisture protection in the form of an exterior wall infiltration barrier must be provided.

Masonry veneer anchored to exterior wall framing must be attached with corrosion-resistant ties of strand wire and be not less than 0.148-in (3.76-mm) (No. 9 gauge) wire with the ends of the wire bent to a 90° (1.6 rad) angle to form a hook not less than 2 in (51 mm) long. The wire ties embedded in the mortar joint must be a minimum of one-half the veneer thickness. Each metal tie must support not more than 3 sq ft (0.28 m^2) of wall area with a maximum spacing of 16 in (406 mm) vertically and 32 in (813 mm) horizontally.

When anchored veneer is applied over framing, the studs have an on-center maximum of 16 in (406 mm) horizontally and are faced with sheathing materials as specified in the local adopted code on both sides. The minimum thickness of the sheathing material or insulation board installed on each side of the framing is $^1/_2$ in (12.7 mm). A 1-in (25.4-mm) minimum air space must be maintained between the anchored veneer and the exterior face of the sheathing material or insulation board. Moisture protection is also required.

Exterior Trim

Architectural trim installed on the framing structure of Types I, II, and IV construction may be of Type VI construction, and on all buildings of Type III, V, and VI construction may be of Type VI construction. Typical residential outlooker and cantilever trim as shown in Figs. 4-22 to 4-24 must be secured to the exterior wall with metal or other approved brackets or fasteners. When architectural trim is located along the top of exterior walls, it must be completely backed by the exterior wall and cannot extend over the top of exterior walls.

Decks not used as exterior exits and bay windows that project from exterior walls must conform to the type of construction required for the building to which they are attached, except that exterior fire-retardant-treated wood is permitted on buildings three stories or less for Type I and II exterior walls.

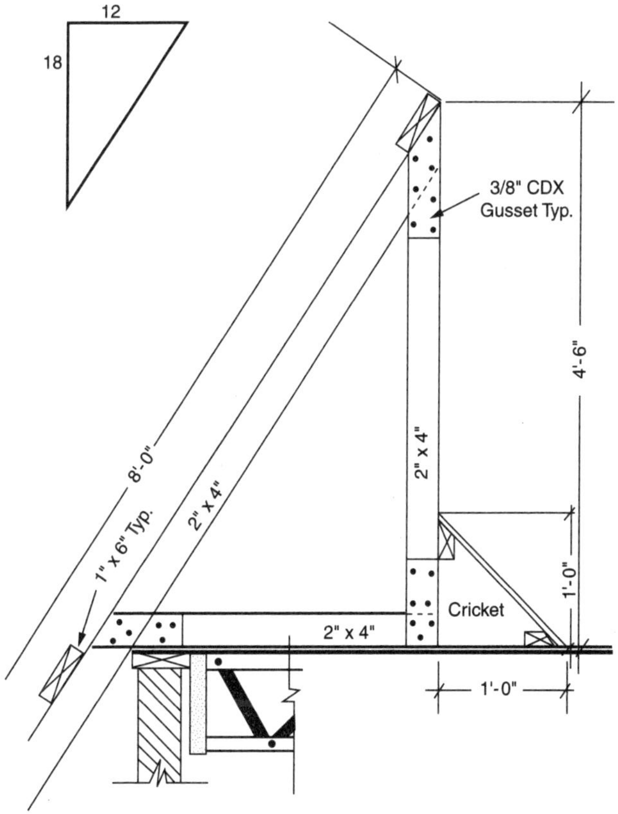

Figure 4-22 Outlooker Framing

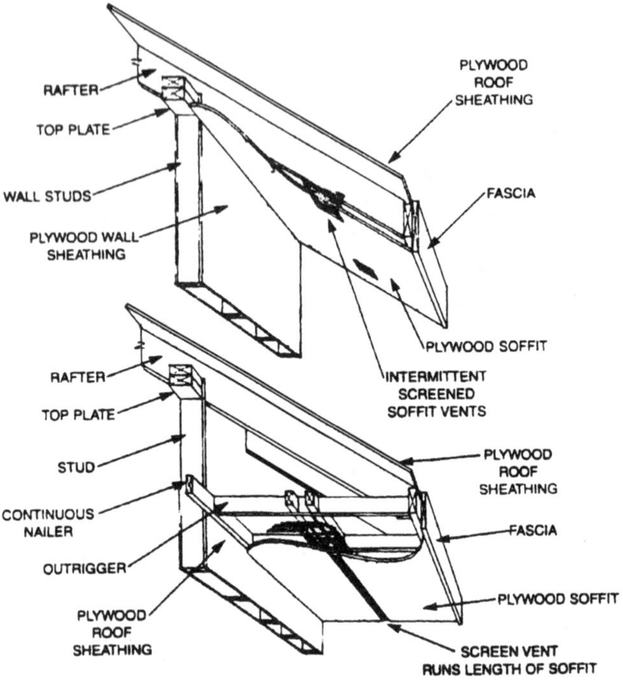

Figure 4-23 Outlooker Trim

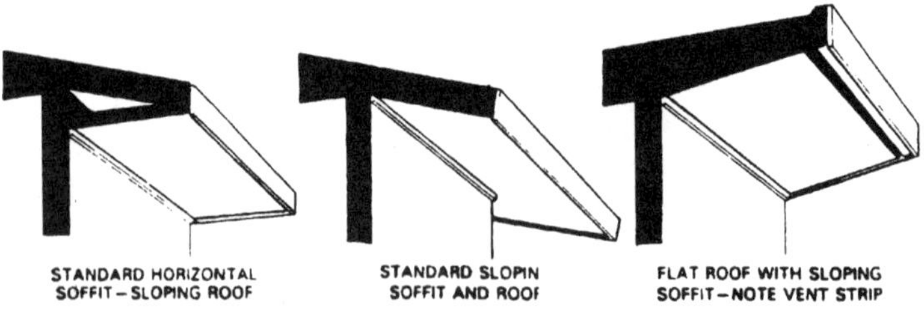

Figure 4-24 Cantilever Trim

Combustible projections from exterior walls located where protection of openings is required must be 1-hour fire-resistance or heavy timber construction. Projections cannot extend more than 12 in (305 mm) into the areas where openings are prohibited.

Fire Department Access in Exterior Walls

Exterior walls must have access openings for fire department use serving each story above grade on an accessible side of the building up to a height of 75 ft (22.9 m). Such access openings must be a minimum of 32 in (813 mm) wide and 48 in (1219 mm) high and with the bottom of the opening not more than 32 in (813 mm) above the floor. Openings must be so spaced that there will be one opening in each 50 ft (15.2 m) of exterior wall on an accessible side of the building.

The exception here is that residences equipped with a National Fire Protection Association (NFiPA) approved automatic sprinkler system throughout in accordance with NFiPA 13 must have access panels as set forth for each 200 ft (61 m) of exterior wall. Where complying access openings are not apparent, they must have distinctive markings for identification.

Exterior Wall Infiltration Barriers

Because the building codes and the modern residential home buyer are demanding more energy efficiency from houses today, the exterior wall component systems should include a vapor-air infiltration barrier (commonly known in the trades as a "housewrap"). Exterior wall infiltration barriers are being adopted by the building codes because of their proven energy efficiency and are frequently installed in the framing stage.

Wrapping the house in a sheath of new technology fabric is expensive, but the accrued savings in annual heating and air-conditioning energy costs make this step in modern exterior wall construction a money-making investment to home buyers, if the builder just takes time to point it out to them.

Housewraps provide a vapor-air infiltration barrier and exfiltration retarder for the frame of the house. They basically keep conditioned air from leaking out and external air from seeping in. They also allow water vapor to escape, preventing moisture buildup that can rot exterior walls. Plus, it covers construction gaps and seams left by joints and corners. In many areas around the country, an approved housewrap has become a code requirement.

Vapor-air infiltration barriers are most definitely a major selling point feature that knowledgeable home buyers are looking for from modern builders, especially if housewraps are used in combination with an insulating band installed at the framing to foundation sill as shown in Fig. 4-25.

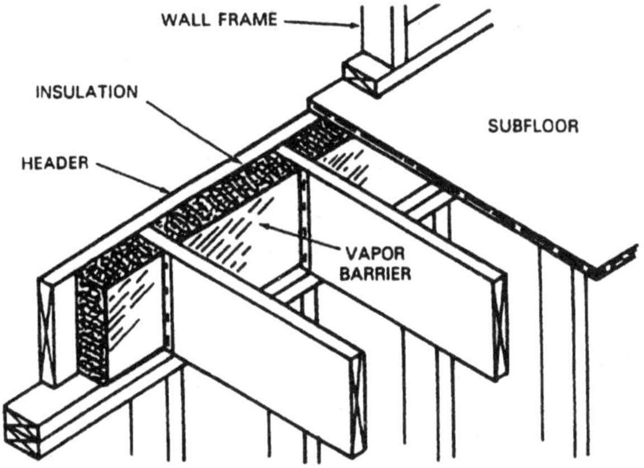

Figure 4-25 Insulating Band

Several brands of exterior wall infiltration barriers are on the market currently. Price varies, but what is really important for the builder is the strength of the wrap. If the product tears easily, it will lose some or all of its barrier capabilities. As usual for the builder, quality of the product equates to quality of the job. Here are the specs you should call for in a quality housewrap used by the manufacturer:

- Trapezoidal tear strength (prior to exposure): 200 percent
- Tear strength after exposure (0 weeks' exposure): 100 percent
- 8 weeks' exposure: 90 percent
- 16 weeks' exposure: 50 percent

- Air porosity (percentage of air stopped): >95 percent
- Moisture vapor transmission: >35 g/m^2 per 24 hours
- Ultraviolet resistance (exposure time before product must be covered): 120 days

By code requirements, homes are constructed to be energy-efficient. Homes built for areas of cold winters have code standard framing sized walls constructed of 2 × 6 in studs with full exterior wall insulation. Post-and-beam framing details are shown in Fig. 4-26. Roof systems use a minimum of 12 in blown cellulose insulation with a minimum R value of 38. Other new energy-saving techniques include seamless housewraps for corners and interior wall plates, which provide moisture control and roof insulation.

The building process also allows for installation of the vapor barrier in the ceiling before the interior walls are in place. This way, the vapor barrier is virtually continuous and not broken by interior wall plates.

Additions to Existing Buildings

An addition that is structurally independent from an existing building frame must be designed and constructed in accordance with the seismic requirements for new buildings. An addition that is not structurally independent from an existing building is designed and constructed such that

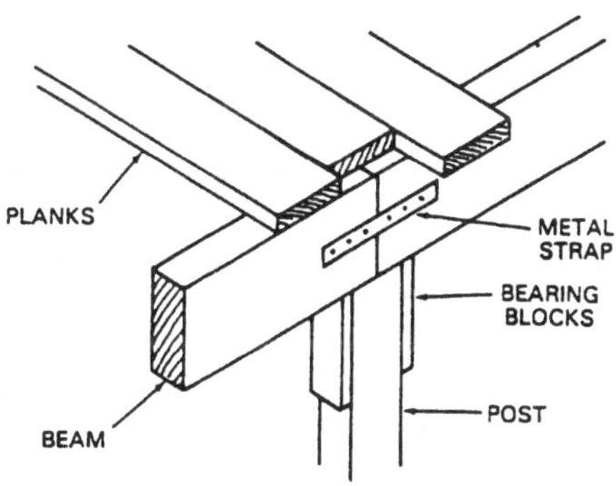

Figure 4-26 Post-and-Beam Details

the entire building conforms to the seismic requirements for new buildings unless the following three provisions are complied with:

- The addition complies with the seismic requirements for new buildings.
- The addition does not increase the seismic forces in any structural element of the existing building by more than 5 percent unless the increased forces on the element are still in compliance with these provisions
- The addition does not decrease the seismic resistance of any structural element of the existing building below that required for new buildings

When a change of structural framing results in an existing building being reclassified to a higher seismic hazard exposure group, the building must conform to the seismic requirements for new buildings. The new exterior walls must also improve the delta factor heat and cooling energy loss factors as shown in Fig. 4-27.

Panelized Framing Construction

In panelized framing construction, new materials such as engineered wood products and noncombustible structural

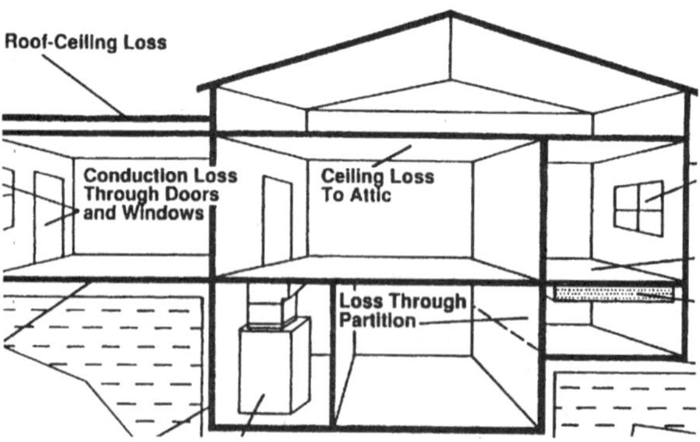

Figure 4-27 Delta Factors

fiber-reinforced cement board are used as shear wall sheathing and exterior siding. It is comprised of portland cement, mineralized cellulose fibers, and mineral additives. The cellulose fiber contained in cement board is derived from recycled materials.

Noncombustible structural fiber-reinforced cement board is a safe and cost-effective alternative to traditional exterior wall plywood problems. It reduces fire sprinkler requirements and allows lower insurance costs and fire-resistant design. Select exterior wall and deck sections in combination with cement board thicknesses for load carrying capacity over given framing spacing.

Limit deflection under design load to 1/240 maximum deflection is possible with cement board. Use backing in all exterior wall and floor assemblies where spans are greater than 24 in on center. Noncombustible structural fiber-reinforced cement board allows for construction of noncombustible wall and roof designs. In panelized construction, it offers the durability of concrete without the time delays and equipment expense related to pouring wet concrete.

Noncombustible structural fiber-reinforced cement board has proved itself to be an excellent choice as an exterior wall sheathing behind many types of wall finishes. Because of its composition of silicone impregnated portland cement, it makes a highly cost effective product that will still perform to design specifications long after plywood or gypsum-based sheathings have failed. Structural fiber-reinforced cement board is also highly impact-resistant and is a perfect choice as an interior wall panel for kitchens and bathrooms and high-traffic areas where excessive use is a concern.

The load-carrying capabilities of structural fiber-reinforced cement board allow its use as a floor framing component. Structural fiber-reinforced cement board has proved itself to be a cost-effective alternative to plywood when a noncombustible, rot-proof, and vermin-proof product is required, and a welcome substitute for costly metal deck and poured concrete flooring. Its high quality and ability to be easily installed and its unique “clean cut” make it well-suited for modern residential construction.

Noncombustible cement board is also used as a sheathing over roof decks, supported by trusses spaced 4 ft on center. This has proved itself to be a working combination when a noncombustible roof assembly is required. The ability to staple-attach asphalt roof shingles to this cement board creates an aesthetically pleasing look that many architects desire. The National Fire Protection Association also states that when a noncombustible roof assembly is incorporated into the overall building design, the residence will probably qualify for a reduction of fire sprinkler requirements.

Portland cement building panels have passed the requirements of ASTM E136 and are rated for Zero Flame Spread and Zero Smoke Developed under ASTM E84. They can be installed as a roof sheathing in lieu of plywood to provide a totally noncombustible roof assembly. Depending on local building codes, sprinkler requirements may be able to be reduced due to the noncombustible roof design. Additional savings could possibly be attained by lower insurance costs as well. Noncombustible structural fiber-reinforced cement board has been tested and approved under U.S. Testing Lab No. 107681, for the staple attachment of asphalt shingles.

This meets code requirements as a structural sheathing and can be used to construct exterior walls and floors, in some cases eliminating the need for metal deck and concrete, and is ideally suited for use as a subfloor over light framing. Tested as a component of a loadbearing floor and ceiling assembly, noncombustible structural fiber-reinforced cement board attains a $1^1/_2$-hour fire rating under ASTM E119-88.

Accordingly, it can be installed as exterior wall sheathing behind brick or exterior wall panels and as a base for exterior synthetic finishes. This product, being silicone-impregnated, will also resist surface water penetration. It is easily cut and installed using standard carpentry tools and fasteners. It can also be routered to produce reveals and synthetic finish applied to create a truly unique appearance.

Light gray in color, standard production is one smooth side and one rougher side. Calibrated noncombustible structural fiber-reinforced cement board is available by special order S2S (sanded smooth both sides). Finishes are usually, but not necessarily, applied to the smooth side of boards, although its surface is not intended as a wear surface.

Plycem board is manufactured and sold as a structural substrate sheathing material intended to be used as a component in roof, floor, and wall assemblies. Standard production is square-edged board. Some special edge shapes are possible on special orders. Standard production is 4 × 8 ft and 4 × 10 ft. Maximum width available is 4 ft and maximum length is 10 ft. Widths and lengths less than maximum can be custom-produced for 500-piece minimum orders. Standard stocked thicknesses are metric: 8 mm. ($^5/_{16}$ in), 11 mm ($^7/_{16}$ in), 17 mm ($^5/_8$ in), and 22 mm ($^7/_8$ in). Additional special order thicknesses available are 6 mm ($^1/_4$ in), 20 mm ($^3/_4$ in), 25 mm (1 in), and 30 mm. ($1^3/_{16}$ in).

When transporting loose boards of structural fiber-reinforced cement board by truck, they must be laid flat, fully protected against edge damage, and covered with waterproof covering. When hand carrying single boards, they must be carried on edge with the 4 ft board dimension vertically. Boards should be stored flat on level supports not exceeding 32 in on centers. Never store on edges or upright. If stored outdoors, a protective covering must be secured over the pallets. If not in the original factory crate, do not stack more than 100 boards high.

Although structural fiber-reinforced cement board is silicone-impregnated to repel surface water absorption, it remains vapor permeable to airborne humidity and other vapors. Extreme variances of humidity and temperature will manifest slight dimensional changes that should be considered in the design of the intended application. It is recommended to design joints in panelized construction (Fig. 4-28) that will allow for movement.

They come from the factory with a moisture content of 8 percent (± 3 percent) and are in equilibrium when temperature is 68°F (20°C) with relative humidity of 50 to 60 percent. The boards remain water vapor permeable and will adapt to the ambient humidity level. Plycem must be allowed to acclimatize for 24 to 48 hours before installation

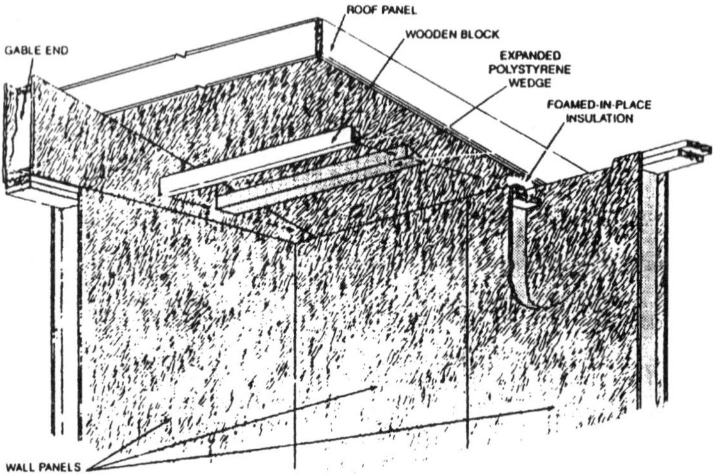

Figure 4-28 Panelized Construction

in the working environment of its intended use. Never install structural fiber-reinforced cement board while the boards are wet or damp.

Although this stuff has the density of a brick, it is easily cut and installed using standard carpentry tools. Use ordinary crosscut hand saws, jig saws with coarse blades, or circular power saws with 18-tooth cross-cut blades. Cut in well-ventilated area, as it contains portland cement. It can easily be nailed or screwed to support framing members. A support backing must always occur behind the fastener location.

Do not nail or screw collateral exterior wall building materials to structural fiber-reinforced cement board without a secure backing surface behind the board (to prevent spalling of back side). It has a pH value of approximately 10 and consists primarily of portland cement, so screws and nails used to attach to supports must be treated for corrosion resistance. Do not use phosphate-treated screws.

Maintain maximum spacing along supports of 12 in o.c. with all fasteners. Centerline of fasteners is to maintain a minimum distance from board edges of $^3/_4$ in. Do not overdrive screw heads. Seat screw heads flush with board surface. Use variable-speed, 2500 rpm maximum, electric screwdrivers. Maintain a minimum of 2-in distance from all board corners with any fastener. Do not locate fasteners at 45° from each other at board corners.

Production of panelized construction is done on jig tables on which the manufacturing plant puts together the closed wall panels. Steel in the panels typically ranges from 25 to 14 gauge. Headers are back-to-back C-channels. Self-tapping screws are driven by coil-fed electric or pneumatic tools. The reason panelized construction makes use of more steel orientation than wood is that closed wood panels erected in a wall sometimes tend to grow due to swelling of the wood.

In a 35-ft-long wall, swelling as much as $^3/_4$ in. in expansion from one end of the wall to the other can happen in areas of high temperature swings. This leads to significant problems with both the electrical and plumbing systems.

By contrast, in a 35-ft-long panelized framed wall the expansion is only about $^3/_{16}$ in from one end to the other. The most common cause of the higher expansion in the wood is use of green lumber, a common supply problem in the southwest, in addition to the area's hot-cold climate conditions. Climate has a slight effect on shrinking or expanding panelized framed walls.

Exterior wall panels can be designed to include electrical and plumbing lines preinstalled, complete with exterior siding and interior drywall, taped, textured, and painted. The relatively small 4 × 8 panel sizes represent the standard modulus of construction, and in many areas where the homes are to be shipped mechanical lifting equipment is not available, but there seems to be plenty of manual labor. Two men can quite easily handle 4 × 8 panels in construction.

Also, because the framework is doubled at each side of the 4 × 8 panels, the framer ends up with a stronger frame. On some homes, such as three-story designs, 4 × 8 shear panels made of $^1/_4$ in panelized construction are placed in strategic areas to give the structural framing engineering validity.

Structural trusses fit into saddles on the top of the C-channel. Two grade A bolts are then inserted through the saddle and truss. The trusses are placed at 4-ft centers. The panelized truss utilizes 2 × 4 in structural members. These members are fastened together into wall and roof panels, trusses, and headers. The engineered components are fabricated within a precise tolerance. The structural wall panels are designed to accommodate attachment of exterior materials on standard framing 16-in centers. C-channels are located on the interior side of the panels so interior wallboard can be attached at 24-in centers.

Framing crews first attach the wall or roof panel to the trusses. They start at one end of the structure and connect the panels at the crest of the trusses, first one side and then the other, working their way toward the other end of the framing structure. The crew then connects together two halves of a scissors truss that will provide vaulted ceiling rafters.

For homes located in a 70 mph wind exposure C-zone or more, special engineering must go into the framing design. Here the panelized construction framing system must be engineered and manufactured for a structure built to withstand 100 mph wind exposure D-zone criteria. A building shell constructed with this type of panelized construction can be a clear-span perimeter system, so all interior partitioning is nonloadbearing.

Panelized wall systems allow for plenty of design flexibility. Standard trusses are now designed to clear-span up to 48 ft and to accommodate a $^{4}/_{12}$ roof pitch. The trusses fit into saddles in the top of the framed columns for placement at 4 ft centers. The trusses and wall panels are connected at each saddle. Scissors trusses for the vaulted ceilings can be delivered in two parts and connected together at the site.

After the crew attaches the trusses to the wall panels, roof panels are connected to the trusses. The framing crew starts at one end of the structure and attaches the panels at the crest of the trusses, first one side and then the other, working its way toward the other end of the structure. Then the framing crew works its way back across, attaching the panels to the lower end of the trusses.

Once assembled, panelized walls connected to the structural framing systems absorb all live and dead loads, and outside shear and seismic forces. The framing shell is at this point strong and durable enough that exterior walls are full loadbearing and interior partitions are nonloadbearing.

Panelized construction includes roofing as well as walls. Fiberglass and tile panel roofing combine the popular look of Spanish tile with the best features of metal panels, including light weight, durability, and all-weather performance. Proven wind resistance of 230+ mph is specified for the best-quality stuff. Systems are typically made up of 3-ft-wide panels in lengths up to 20 ft, with a full range of accessories designed for builder-friendly installation. These are available in a wide range of traditional tile styles and designer colors. Some even look just like granite slate. You can't tell the difference standing in the driveway looking at them. Cool stuff indeed, and no need for an expensive 2 × 6 rafter system to carry the heavy load of the real thing.

Panelized walls use exterior steel doors to combine the beauty of wood with the security of 22-gauge steel. A special primer coat seals the walls and accepts most stains to give the look and feel of natural wood. Recessed hinge systems allow for alignment adjustments of exterior doors in their frames.

Header Code Design Charts

The following header code structural design charts (Tables 4-7 and Table 4-8) show nominal sizing and location of framing header assemblies and where they are located.

Table 4-7 Header Design Chart

Header Design Chart
(Spruce-Pine-Fir No. 2, Roof Load = 20LL + 10DL, Floor Load = 40LL + 10DL)
$F_b = 875$ $E = 1{,}300{,}000$ $F_v = 70$

Header Supporting:	Header Size:	Maximum Allowable Header Space (ft-in) Nonstructural Sheath. House Depth (ft)			1/2" Insul. Bd. Sheath. House Depth (ft)			15/32" Plywood Sheath. House Depth (ft)		
		24	28	32	24	28	32	24	28	32
Roof	2-2x4	4-2	3-11	3-8	4-5	4-3	4-0	5-2	5-0	4-9
	2-2x6	5-10	5-5	5-1	6-2	5-9	5-5	7-1	6-7	6-3
	2-2x8	7-4	6-10	6-6	7-8	7-2	6-9	8-6	7-11	7-6
	2-2x10	9-2	8-7	8-1	9-5	8-9	8-3	9-10	9-6	8-11
	2-2x12	10-9	10-3	9-8	10-10	10-6	9-11	11-3	10-10	10-6
Roof + One Story (Bearing)	2-2x4	4-10	4-7	4-4	5-0	4-9	4-6	5-4	5-1	4-10
	2-2x6	5-10	5-6	5-2	6-1	5-8	5-4	6-8	6-3	5-10
	2-2x8	6-10	6-4	6-0	7-1	6-7	6-2	7-7	7-1	6-8
	2-2x10	8-1	7-6	7-1	8-3	7-8	7-2	8-9	8-2	7-8
	2-2x12	9-4	8-8	8-2	9-6	8-10	8-4	10-0	9-4	8-9
Roof + One Story (No Bearing)	2-2x4	4-3	3-11	3-8	4-5	4-1	3-10	4-9	4-6	4-3
	2-2x6	5-0	4-7	4-4	5-2	4-9	4-5	5-8	5-3	4-11
	2-2x8	5-9	5-4	5-0	6-0	5-6	5-2	6-5	6-0	5-7
	2-2x10	6-10	6-4	5-11	6-11	6-5	6-1	7-5	6-10	6-5
	2-2x12	7-11	7-4	6-11	8-1	7-6	7-0	8-6	7-11	7-5
Roof + Two Story (Bearing)	2-2x4	4-3	3-11	3-8	4-5	4-1	3-10	4-9	4-6	4-3
	2-2x6	5-0	4-7	4-4	5-2	4-9	4-5	5-8	5-3	4-11
	2-2x8	5-9	5-4	5-0	6-0	5-6	5-2	6-5	6-0	5-7
	2-2x10	6-10	6-4	5-11	6-11	6-5	6-1	7-5	6-10	6-5
	2-2x12	7-11	7-4	6-11	8-1	7-6	7-0	8-6	7-11	7-5
Roof + Two Story (No Bearing)	2-2x4	3-4	3-1	2-11	3-6	3-3	3-0	3-10	3-7	3-4
	2-2x6	4-0	3-8	3-5	4-1	3-10	3-7	4-6	4-2	3-11
	2-2x8	4-7	4-3	4-0	4-9	4-5	4-2	5-1	4-9	4-5
	2-2x10	5-5	5-1	4-9	5-7	5-2	4-10	5-11	5-6	5-2
	2-2x12	6-4	5-10	5-6	6-5	6-0	5-7	6-9	6-3	5-11

Table 4-7 Header Design Chart *(Continued)*

Header Design Chart
(Hem Fir No. 2, Roof Load = 20LL + 10DL, Floor Load = 40LL + 10DL)
F_b = 1000 E = 1,400,000 F_v =75

Header Supporting:	Header Size:	Nonstructural Sheath. House Depth (ft)			1/2" Insul. Bd. Sheath. House Depth (ft)			15/32" Plywood Sheath. House Depth (ft)			
		Maximum Allowable Header Space (ft-in)									
		24	28	32	24	28	32	24	28	32	
Roof	2-2x4	4-5	4-2	3-11	4-8	4-5	4-3	5-4	5-1	4-11	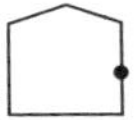
	2-2x6	6-2	5-9	5-5	6-7	6-1	5-9	7-4	7-1	6-8	
	2-2x8	7-10	7-4	6-11	8-3	7-8	7-3	8-7	8-4	8-0	
	2-2x10	9-6	9-2	8-8	9-8	9-4	8-10	10-0	9-8	9-5	
	2-2x12	10-11	10-7	10-3	11-1	10-8	10-5	11-5	11-1	10-9	
Roof + One Story (Bearing)	2-2x4	4-12	4-9	4-6	5-1	4-10	4-8	5-5	5-2	5-0	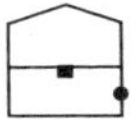
	2-2x6	6-3	5-10	5-5	6-6	6-1	5-8	7-0	6-8	6-3	
	2-2x8	7-4	6-10	6-4	7-6	7-0	6-7	8-1	7-7	7-1	
	2-2x10	8-7	8-0	7-6	8-10	8-2	7-8	9-4	8-8	8-2	
	2-2x12	10-0	9-4	8-9	10-2	9-6	8-11	10-5	10-0	9-4	
Roof + One Story (No Bearing)	2-2x4	4-5	4-2	3-11	4-7	4-4	4-1	4-10	4-8	4-5	
	2-2x6	5-4	4-11	4-7	5-6	5-1	4-9	6-1	5-7	5-3	
	2-2x8	6-2	5-9	5-5	6-4	5-11	5-6	6-10	6-5	6-0	
	2-2x10	7-3	6-9	6-4	7-5	6-11	6-6	7-11	7-4	6-11	
	2-2x12	8-5	7-10	7-4	8-7	8-0	7-6	9-1	8-5	7-11	
Roof + Two Stories (Bearing)	2-2x4	4-5	4-2	3-11	4-7	4-4	4-1	4-10	4-8	4-5	
	2-2x6	5-4	4-11	4-7	5-6	5-1	4-9	6-1	5-7	5-3	
	2-2x8	6-2	5-9	5-5	6-4	5-11	5-6	6-10	6-5	6-0	
	2-2x10	7-3	6-9	6-4	7-5	6-11	6-6	7-11	7-4	6-11	
	2-2x12	8-5	7-10	7-4	8-7	8-0	7-6	9-1	8-5	7-11	
Roof + Two Stories (No Bearing)	2-2x4	3-7	3-4	3-1	3-9	3-6	3-3	4-2	3-10	3-7	
	2-2x6	4-3	3-11	3-8	4-5	4-1	3-10	4-10	4-6	4-2	
	2-2x8	4-11	4-7	4-3	5-1	4-9	4-5	5-6	5-1	4-9	
	2-2x10	5-10	5-5	5-1	5-11	5-6	5-2	6-4	5-10	5-6	
	2-2x12	6-9	6-3	5-10	6-10	6-4	6-0	7-3	6-8	6-3	

1 in = 25.4 mm   1 ft = 0.305 m   1 psi = 6.8948 kPa

Table 4-8 Maximum Allowable Header Span

Maximum Allowable Header Span (ft-in)
Design Procedure: Single Top Plate Construction: Interior Bearing Walls

Species, Grade Properties	Header Supporting	Header Size	Nonstructural Sheath. House Depth (ft)			1/2" Insul. Bd. Sheath.[2] House Depth (ft)			15/32" Plywood Sheath.[3] House Depth (ft)		
			24	28	32	24	28	32	24	28	32
Douglas Fir Larch Southern Pine No. 2 E = 1,600,000 F_b = 1200 F_v = 90	One Story Bearing	2-2x4	3-5	3-2	3-0	3-10	3-7	3-4	4-7	4-4	4-2
		2-2x6	5-1	4-8	4-5	5-5	5-0	4-8	6-4	5-10	5-6
		2-2x8	6-7	6-1	5-8	6-10	6-4	5-11	7-8	7-1	6-7
		2-2x10	8-3	7-8	7-2	8-6	7-11	7-4	9-1	8-6	8-0
		2-2x12	10-0	9-3	8-8	10-1	9-6	8-10	10-5	10-0	9-5
	Two Story Bearing	2-2x4	2-7	2-4	2-3	2-10	2-8	2-6	3-7	3-4	3-1
		2-2x6	3-9	3-6	3-3	4-0	3-9	3-6	4-8	4-4	4-1
		2-2x8	4-10	4-6	4-2	5-1	4-8	4-5	5-8	5-3	4-11
		2-2x10	6-1	5-8	5-4	6-4	5-10	5-6	6-10	6-4	5-11
		2-2x12	7-5	6-10	6-5	7-7	7-0	6-7	8-1	7-6	7-0
Hem Fir No. 2 E = 1,400,000 F_b = 1000 F_v = 75	One Story Bearing	2-2x4	3-2	2-11	2-9	3-6	3-3	3-0	4-4	4-1	3-10
		2-2x6	4-8	4-3	4-0	4-11	4-7	4-3	5-9	5-4	5-0
		2-2x8	6-0	5-6	5-2	6-3	5-9	5-5	7-0	6-5	6-0
		2-2x10	7-6	7-0	6-6	7-9	7-2	6-9	8-5	7-9	7-3
		2-2x12	9-1	8-5	7-11	9-4	8-8	8-1	10-0	9-3	8-7
	Two Story Bearing	2-2x4	2-4	2-2	2-0	2-7	2-5	2-3	3-3	2-11	2-8
		2-2x6	3-5	3-2	3-0	3-8	3-5	3-2	4-3	4-0	3-9
		2-2x8	4-5	4-1	3-10	4-8	4-3	4-0	5-2	4-9	4-6
		2-2x10	5-7	5-2	4-10	5-9	5-4	5-0	6-3	5-9	5-5
		2-2x12	6-9	6-3	5-10	6-11	6-5	6-0	7-5	6-10	6-5
Spruce Pine Fir No. 2 E = 1,300,000 F_b = 875 F_v = 70	One Story Bearing	2-2x4	2-11	2-9	2-7	3-3	3-0	2-10	4-1	3-10	3-7
		2-2x6	4-4	4-0	3-9	4-8	4-3	4-0	5-5	5-0	4-8
		2-2x8	5-7	5-2	4-10	5-10	5-5	5-1	6-6	6-0	5-8
		2-2x10	7-1	6-6	6-1	7-3	6-9	6-3	7-10	7-3	6-10
		2-2x12	8-6	7-11	7-5	8-9	8-1	7-7	9-4	8-7	8-1
	Two Story Bearing	2-2x4	2-2	2-0	1-11	2-5	2-3	2-1	3-1	2-10	2-6
		2-2x6	3-3	3-0	2-9	3-5	3-2	3-0	4-0	3-9	3-6
		2-2x8	4-2	3-10	3-7	4-4	4-0	3-9	4-10	4-6	4-2
		2-2x10	5-3	4-10	4-6	5-5	5-0	4-8	5-10	5-5	5-1
		2-2x12	6-4	5-10	5-6	6-6	6-0	5-7	6-11	6-5	6-0

1 in = 25.4 mm   1 ft = 0.305 m   1 psi = 6.8948 kPa

Table 4-8 Maximum Allowable Header Span *(Continued)*

Maximum Allowable Header Span (ft-in)[1]

Design Procedure: Double Top Plate Construction: Interior Bearing Walls

Species, Grade Properties	Header Supporting	Header Size	Nonstructural Sheath. House Depth (ft)			1/2" Insul. Bd. Sheath.[2] House Depth (ft)			15/32" Plywood Sheath.[3] House Depth (ft)		
			24	28	32	24	28	32	24	28	32
Spruce	One Story	2-2x4	3-3	3-0	2-10	3-7	3-4	3-1	4-4	4-0	3-9
Pine	Bearing	2-2x6	4-7	4-2	3-11	4-10	4-6	4-2	5-7	5-2	4-10
Fir No. 2		2-2x8	5-9	5-4	5-0	6-0	5-7	5-2	6-8	6-2	5-9
E = 1,300,000		2-2x10	7-2	6-8	6-3	7-5	6-10	6-5	8-0	7-5	6-11
F_b = 875		2-2x12	8-8	8-0	7-6	8-10	8-2	7-8	9-5	8-9	8-2
F_v = 70	Two Story	2-2x4	2-5	2-3	2-1	2-8	2-5	2-3	3-3	3-0	2-9
	Bearing	2-2x6	3-4	3-1	2-11	3-7	3-4	3-1	4-2	3-10	3-7
		2-2x8	4-3	3-11	3-8	4-5	4-1	3-10	4-11	4-7	4-3
		2-2x10	5-4	4-11	4-7	5-6	5-1	4-9	5-11	5-6	5-2
		2-2x12	6-5	5-11	5-7	6-7	6-1	5-8	7-0	6-6	6-1

1 in = 25.4 mm 1 ft = 0.305 m 1 psi = 6.8948 kPa

Notes

1. Linear interpolation for house widths not in table is permitted. For example, assume a 26 ft (7924 mm) wide house with 15/32 inch (11.9 mm) plywood sheathing - 2x6 header allowable header span = 7 ft - 5 in (2261 mm). Header spans identified as having "no bearing" construction apply to both interior and exterior loadbearing walls which have tributary areas equal to one-half the house depth. Header spans identified with "bearing" construction apply only to exterior bearing walls with tributary areas equal to one-fourth of the house depth.
2. Sheathing or combined sheathing/siding having a minimum density of 18 pcf (288.3 kg/m^3)
3. Minimum 15/32 inch (11.9 mm) plywood sheathing or combined sheathing/siding applied between bottom of header, the top of the top plate and between lines of the broken vertical studs at the end of the header and nailed to the header, top plates, cripples and studs - 6 inches (152 mm) o.c. at the edges and 12" (305 mm) o.c. at intermediate framing.

Notes

Chapter

5

Interior Wall Framing

Framing Lumber Imperfections

Most American framing contractors agree that the quality of wall framing lumber is getting worse. Many builders say that framing lumber quality is much worse than it was 10 years ago, and no one says it's getting any better. What are the most common defects in framing lumber deliveries? Wood that's warped, bent, and bowed tops the list by far. Many builders say knots and waned edges are also big problems for interior walls that must be plumb, line, and level. They also blame splits and checks, improper kiln drying, and inconsistent dimensioning for poor construction-grade building lumber.

The most common builder strategy mentioned at job sites for dealing with lousy lumber delivery drops is "send it back and demand credit for returned lumber." Here are some of the other things competitive builders suggest doing with poor wall framing lumber:

- Order more matèrial than you need, pick through it, and return unusable stock to suppliers. (Some builders eat the extra time this takes; some withhold payment to cover the cost of culling through the load.)

- Require framers to set aside poor-quality joists to use as headers
- Specify better-grade lumber to increase percentage of usable stock in each delivery
- Get suppliers to shop around for better mills
- Use more engineered wood products like I-beams, joists, LVL headers, glue-laminated beams, and finger-jointed studs in estimating takeoffs for component symbols (like those shown in Fig. 5-1)

Why is it so hard to get decent lumber for building a plumb, line, and level interior wall? Most builders blame the federal government for the decline in lumber quality, saying limited harvests have hurt the industry. But many builders also point the finger at the wood products companies. They suspect the wood products companies are pushing wood

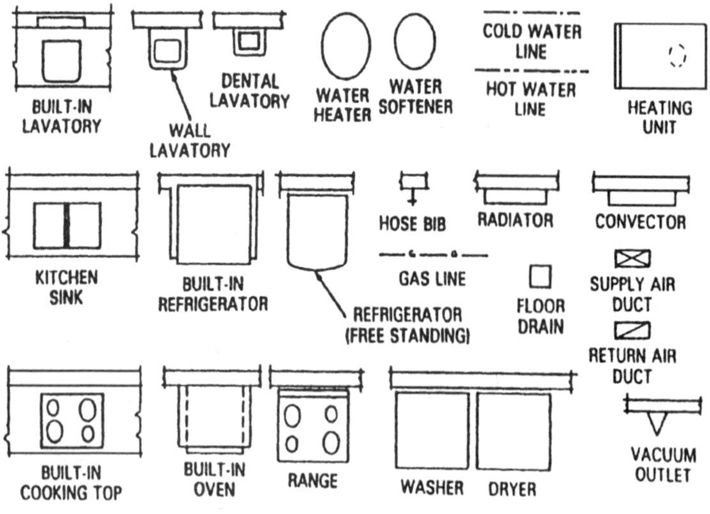

Figure 5-1 Engineered Components

engineered products (therefore neglecting dimension lumber products) and haven't planted enough trees to replenish the resource. The top five defects of modern wall framing lumber are:

1. Warped, bent, and bowed lumber
2. Wood with too many knots; loose knots
3. Board with splits and checks
4. Lumber not properly dried
5. Dimensions not consistent

The number one defense against poor-quality wall framing lumber used by today's framer is to pick it out, send it back, and don't pay for it. Some builders also think the industry exports the best logs to overseas markets, saving the worst for us at home. And some say the problems stem from companies that process smaller, younger trees. Those of us who have been licensed contractors for more than a decade also question how well (if at all) the industry has maintained its grading and drying standards.

Interior Wall Shear

Interior walls built today usually have doubled top plates on which the roof, ceiling, or floor framing above rests. On some older buildings, only a single top plate was installed. The top plate(s) is where the shear wall sheathing is nailed and where the shear forces transfer from the roof, ceiling, or floor diaphragm. For adequate transfer of forces, it is necessary for the framing to attach the top plate to the rim joist or blocking above. Examples of common interior partition walls are shown in Figs. 5-2 and 5-3.

In determining allowable interior wall height (including plates), remember that a roof structure designed to house anything other than mechanical service equipment will be counted as another additional story. For the purpose of plans checking, interior walls are typically shown opposing the exterior walls as shown in Fig. 5-4 and usually include

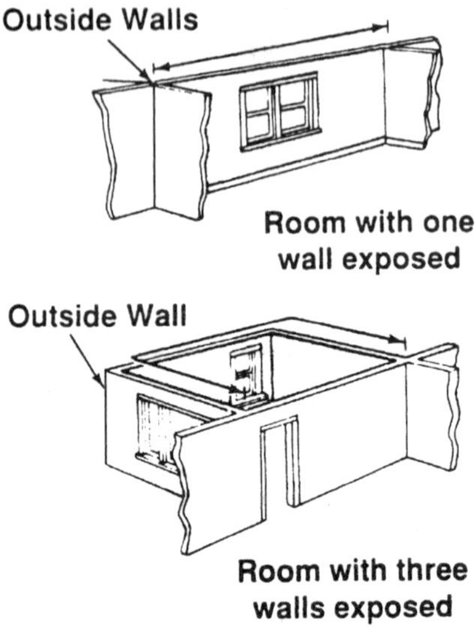

Figure 5-2 Interior Walls

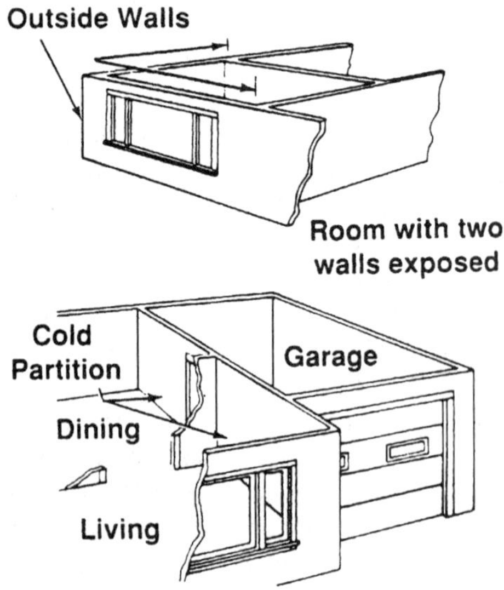

Figure 5-3 Partition Walls

some finish details as shown in Fig. 5-5. Interior walls also differ in insulating R-values for the different heating and cooling loads shown in Figs. 5-6 and 5-7.

Code conventional wall framing requires that each joist be attached to the top plate with three 8d toenails. However, this amount of attachment may not be adequate for shear forces in the plane of the interior wall. Since it is difficult to determine if there is sufficient existing toenailing, metal right-angle clips (framing clips) should be nailed into the top plate(s) and the rim joist or blocking. This will strengthen the shear connection between the floor framing and the shear wall below.

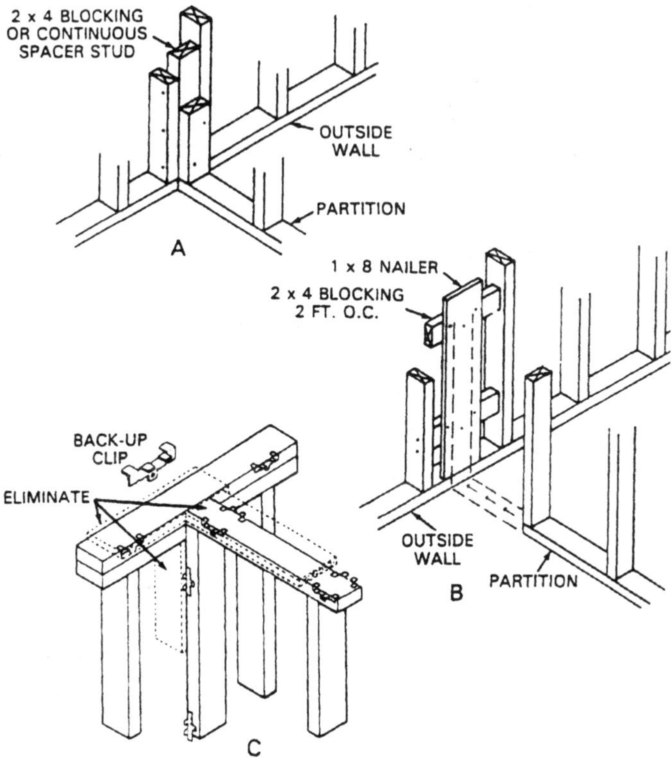

Figure 5-4 Interior & Exterior Wall Inspection

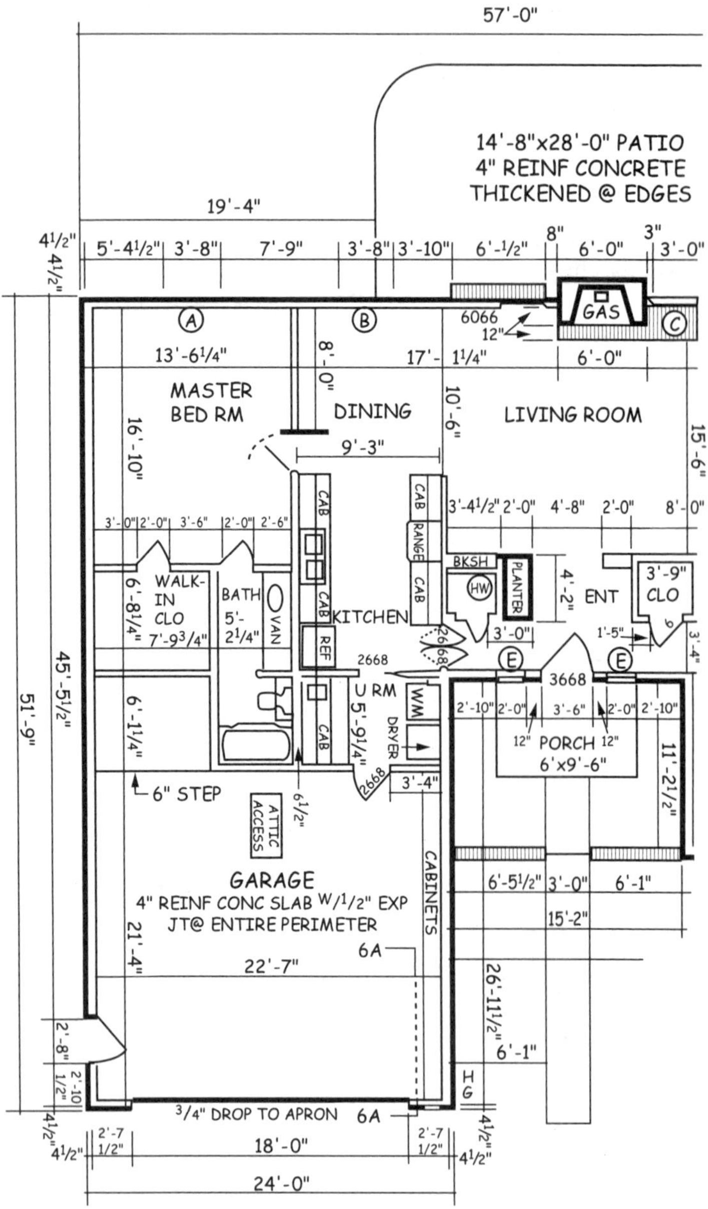

Figure 5-5 Interior Details

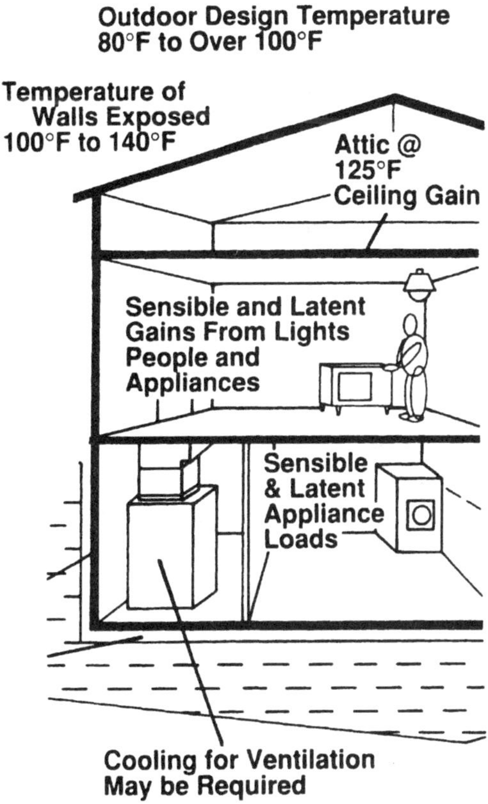

Figure 5-6 Delta Factors

These shear connections should be installed at the required spacing along the entire length of the wall. Spacing will depend on the given design value of the clip and the prescriptive or engineered design that is being used. For shallow 2 × 6 or 2 × 8 joists, it may be difficult to attach the clip into the rim joist or blocking and top plate. The use of a pneumatic palm nailer will make this installation easier in difficult access areas.

Splices in double top plates should be properly lapped and the laps should be well nailed. If a double top plate is

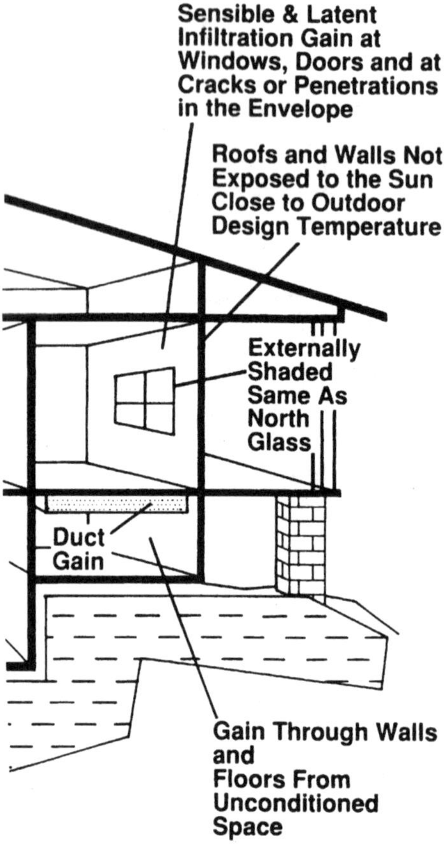

Figure 5-7 Delta Factors

notched or cut (say for a plumbing vent), metal straps should be added across the notch to restore its tension capacity. When splices of double top plate are not properly built or if there is only one discontinuous top plate, a metal strap should be nailed across the splice to provide the necessary continuity.

At the bottom of an interior wall, the attachment of the sole plate and the floor decking to the floor framing accomplishes the shear transfer from the shear wall and the floor diaphragm to the floor framing below. See Fig. 5-8. When nails or lag screws resist shear, they must be sufficiently long to penetrate through the wall's sole plate and floor sheathing into the framing member below. For full load values, Table 23-G of the Uniform Building Code specifies the minimum penetration required of nails into the framing member.

When the nail penetration falls below Table 23-G values, the allowable load for the nail is reduced proportionally up to 50 percent of its value. When nail penetration falls below 6 nail shank diameters, the allowable load value is zero. For this reason, nails often cannot transfer shear forces through thick interior wall sheathing.

If the interior wall framing is exposed above the floor, as in new framing that has not yet had plaster or drywall

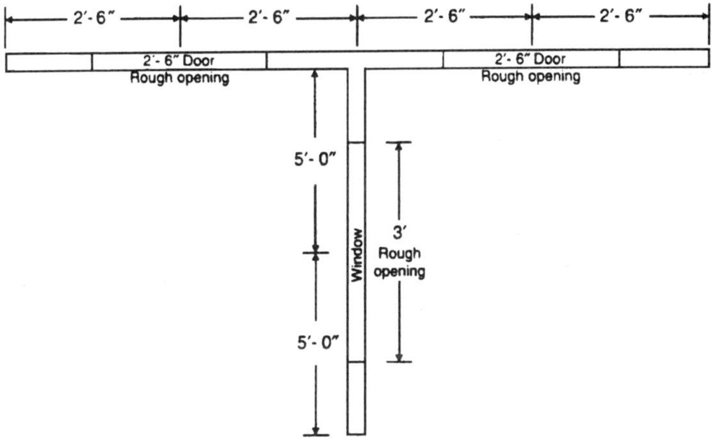

Figure 5-8 Interior Wall Sole Plate

attached, the nails can be installed from above. However, in some situations, the wall framing above is not (or cannot be) exposed. In this case, it may be possible to add framing dips where the underside of the floor and the rim joist or blocking meets. Short joist hanger nails should be used to attach the clips to the joist or blocking. Longer nails or wood screws should be used to penetrate up through the floor decking and into the sole plate above. The interior walls are then completed by room finish schedules as shown in Table 5-1. This is followed by interior doors and windows schedules as shown in Table 5-2.

In some multiple-story framing, double sole plates on the interior may be encountered. The second sole plate may have been installed to act as a form for poured-in-place lightweight concrete or gypsum subfloor material. Where this occurs, it is important to maintain a load path through both plates. In new construction, this is accomplished by nailing the first sole plate and second sole plate with a sufficient number of nails specified by a design professional, then installing baseboard and base shoe (Fig. 5-9). In existing construction, it may be necessary to use extra long nails or screws through both plates so that the fastener has enough penetration into the framing member below.

Another way of accomplishing the shear transfer from an interior wall and floor above into the walls below is to use longer sheets of wood structural panels. By spanning full sheets from the rim joist or blocking above to the rim joist or blocking below, the panel serves as a direct load path from one shear wall to the other. This reduces the number of links in the chain and eliminates the need for framing clips.

Remember to install the four rows of nails (as required by code) to provide a load path from the floor diaphragm as well as the interior wall above and always provide the recommended gap between panel edges to allow for expansion. New construction requires even larger gaps for shrinkage of the rim joist or blocking.

Table 5-1
Finish Schedules

FINISH SCHEDULES
ROOM FINISHES

ROOMS	FLOOR				CEILING				WALL				BASE				TRIM			REMARKS
	CARPET	CERAMIC TILE	RUBBER TILE	CONCRETE	ACOUSTIC TEXT	DRYWALL	PAINT	CERAMIC TILE	DRYWALL	PAINT	WALL PAPER	CERAMIC TILE	WOOD	RUBBER	CERAMIC TILE	STAIN	WOOD	STAIN	PAINT	
ENTRY		√			√				√	√	√		√			√	√	√		See owner for all painting
HALL	√				√				√	√			√			√	√	√		
BEDROOM 1	√				√				√	√	√		√			√	√	√		See owner for grade of carpet
BEDROOM 2	√				√				√	√			√			√	√	√		See owner for grade of carpet
BEDROOM 3	√				√				√	√			√			√	√	√		See owner for grade of carpet
BATH 1	√	√			√			√	√	√	√	√	√			√	√	√		Wall paper three walls around vanity
BATH 2		√			√			√	√	√	√	√			√		√	√		Water seal tile wall paper w. wall
UTIL. & CLOSETS	√		√√			√	√		√√	√			√			√	√√	√	√	Use off-white flat/latex
KITCHEN			√		√				√	√				√			√	√		
DINING	√				√				√	√	√		√			√	√	√		
LIVING	√				√				√	√			√			√	√	√		See owner for carpet
GARBAGE				√		√	√					√		√			√	√		

Table 5-2
Door and Window Schedules

DOOR SCHEDULE

DOOR	WIDTH	HEIGHT	THICKNESS	MATERIAL	TYPE	STORM DOOR	QUANTITY	THRESHOLD	REMARKS	MANUFACTURER
2068	2'-8"	6'-8"	1 3/8"	Wood-Ash	Hollow Core	No	5	None	Oil Stain	L.B.J. Door, Inc.
2468	2'-4"	6'-8"	1 3/8"	Wood-Ash	Hollow Core	No	1	None	Oil Stain	L.B.J. Door, Inc.
2668	2'-6"	6'-8"	1"	Wood-Ash	Cafe	No	1pr.	None	Oil Stain	L.B.J. Door, Inc.
2668	2'-6"	6'-8"	1 3/8"	Wood-Ash	Sliding-Packet	No	1	None	Oil Stain	L.B.J. Door, Inc.
2668	2'-6"	6'-8"	1 3/8"	Wood-Ash	Hollow Core	1 storm	6	1 Alum.	Screen door in garage	L.B.J. Door, Inc.
2868	2'-6"	6'-8"	1 3/4"	Metal Clad	Fire proof	Yes	1	Alum.	Paint	L.B.J. Door, Inc.
3068	3'-0"	6'-8"	1 3/4"	Wood-Ash	Solid Core	No	1	None	Oil Stain	L.B.J. Door, Inc.
3688	3'-6"	6'-8"	1 3/4"	Wood-Ash	Solid Core	Yes	1	Alum.	Norine Varnish	L.B.J. Door, Inc.
6066	6'-0"	6'-6"	1/2"	Glass&Metal	Sliding	Yes	1pr.	Alum.	Sliding Screen	L.B.J. Door, Inc.
6068	6'-0"	6'-8"	1 1/4"	Wood-Ash	Bi Fold	No	25 ets	None	Oil Stain	L.B.J. Door, Inc.
1956	1'-9"	5'-6"	1/2"	Glass&Metal	Slid vs Bi fold door	No	25 ets	None	Frost & Glass	L.B.J. Door, Inc.

WINDOW SCHEDULE

SYMBOL	WIDTH	HEIGHT	MATERIAL	TYPE	SCREEN & STORM	QUANTITY	REMARKS	MANUFACTURER	CATALOG NUMBER
A	3'-8"	3'-0"	Aluminum	Double Hung	Yes	2	4 Lites 4 High	L.B.J. Window Co.	H1 PW
B	3'-8"	5'-0"	Aluminum	Double Hung	Yes	1	4 Lites 4 High	L.B.J. Window Co.	H5 PW
C	3'-0"	5'-0"	Aluminum	Stationary	Storm Only	2	Single Lite	H&J Glass Co.	59 PY
D	2'-0"	3'-0"	Aluminum	Double Hung	Yes	1	4 Lites 4 High	L.B.J. Window Co.	142 PW
E	2'-0"	6'-0"	Aluminum	Stationary	Storm Only	2	20 Lites	H&J Glass Co.	37 TS
F	3'-6"	5'-0"	Aluminum	Double Hung	Yes	1	16 Lites 34 High	L.B.J. Window Co.	147 PW

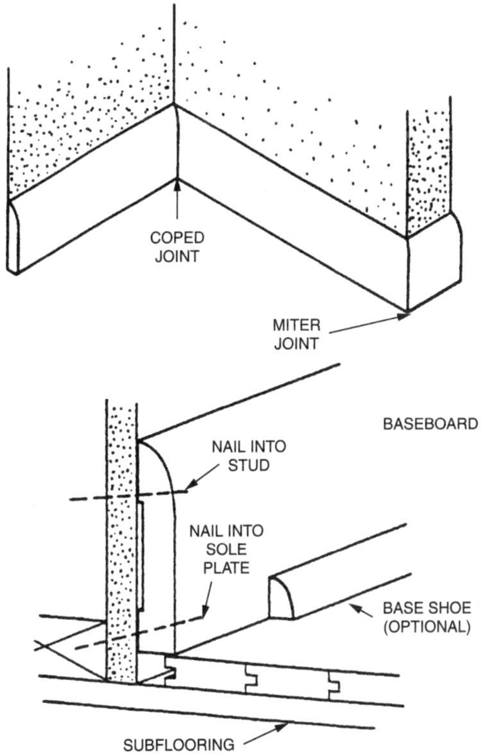

Figure 5-9 Baseboard & Base Shoe

Stud Partition Walls

Stud partition walls containing plumbing, heating, or other pipes must be so framed and the joists underneath so spaced as to give proper clearance for ductwork, piping, and electrical cable routes as shown in Figs. 5-10 and 5-11. Where a partition containing such runs parallel to the floor joists, the joists underneath such partitions must be doubled and spaced to permit the passage of such service components and must be bridged.

Where plumbing, heating, or other pipes are placed in or partly in a partition, necessitating the cutting of sole plates or top plates, a metal tie not less than $^1/_8$ in thick and $1^1/_2$ in wide must be fastened to the plate across and to each side of

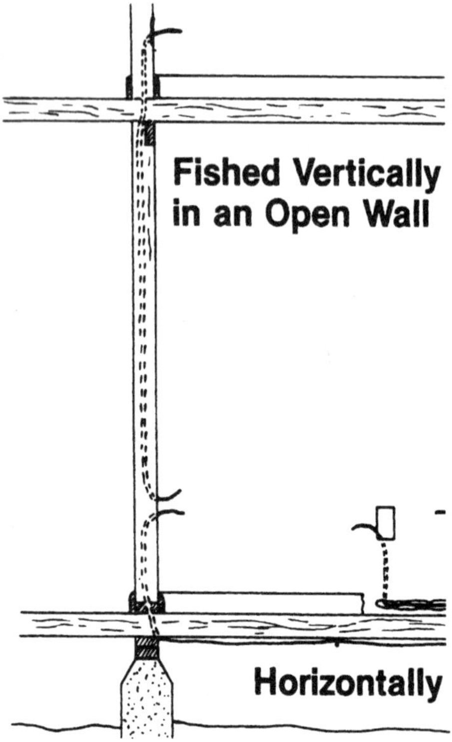

Figure 5-10 Cable Route

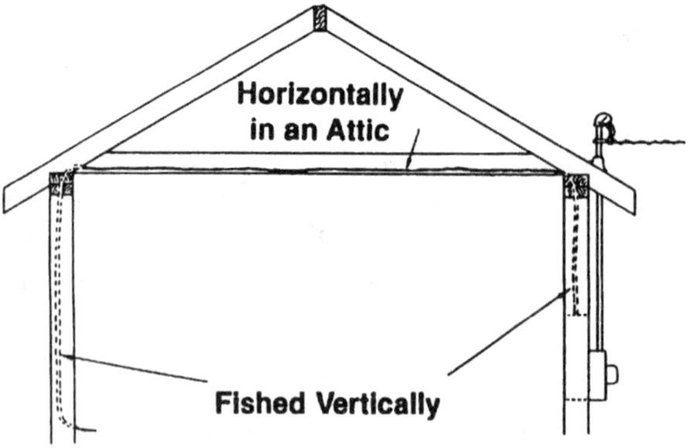

Figure 5-11 Cable Route

the opening with not less than four 16d nails. This is known as *preventive construction*. Preventive construction protects members from sheetrock nails and screws penetrating wiring or piping inside partition walls, especially at framing wall intersections as shown in Fig. 5-12.

A hole not greater than 40 percent of the stud width may be bored in any wood stud. Bored holes not greater than 60 percent of the width of the stud are permitted in nonbearing partition walls or in any wall where each bored stud is doubled, provided not more than two such successive doubled studs are so bored.

The framer must provide firestops in openings around vents, pipes, ducts, chimneys, fireplaces, and similar openings that afford a passage for fire at ceilings and floor levels, with noncombustible materials.

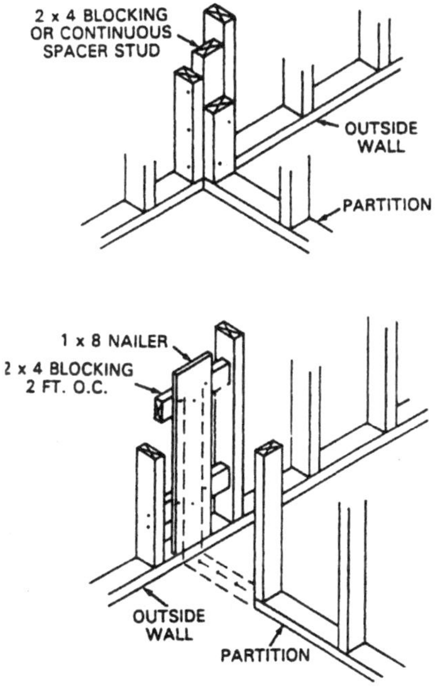

Figure 5-12 Intersecting Wall Framing

All framing in connection with a plumbing system must be so installed that a piping or connections will not be subject to undue strain or stresses, and provisions must be made for expansion and settlement in the structural frame. No structural member can be seriously weakened or impaired by cutting, notching, or boring.

Unless impractical due to structural conditions, all wood beams, girders, joists, studs, and similar construction must be bored with holes approximately the same diameter as the conduit or pipes passing through them. Spans are obtained by taking the information from the different lines in the plans, as shown in Fig. 5-13. Plumbing runs are taken from center dimensions in the interior walls as shown in Fig. 5-14. In multiple story framing, the plumbing stacks go vertically through the interior walls as shown in Fig. 5-15.

In determining allowable building height, remember that a roof structure designed to house anything other than mechanical service equipment will be counted as another additional building story. Upgrade remodeling encounters a few other definitive building code requirements the framer must be aware of.

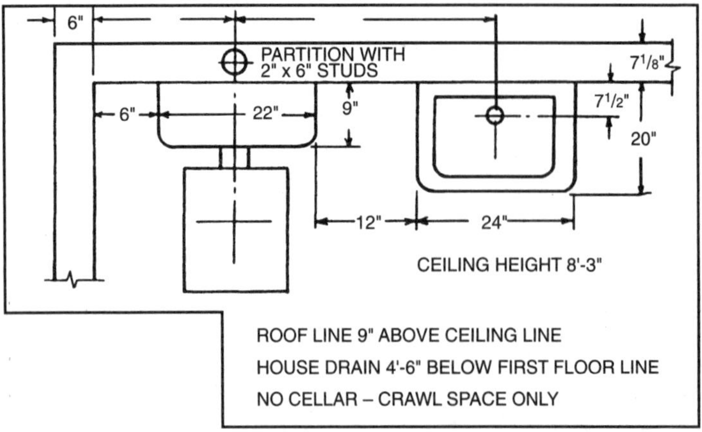

Figure 5-13 Interior Plumbing Layout

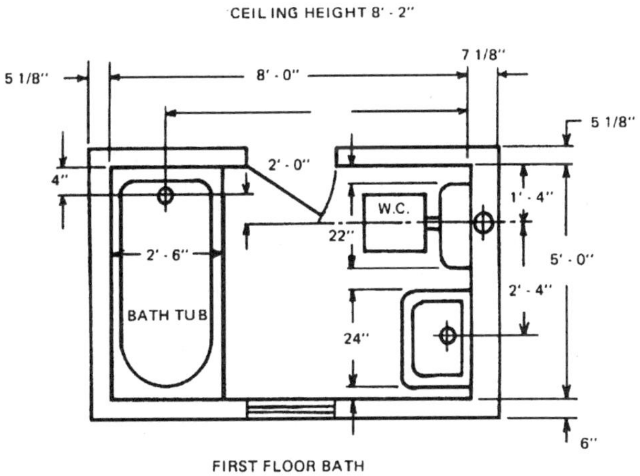

Figure 5-14 Plumbing Dimensions

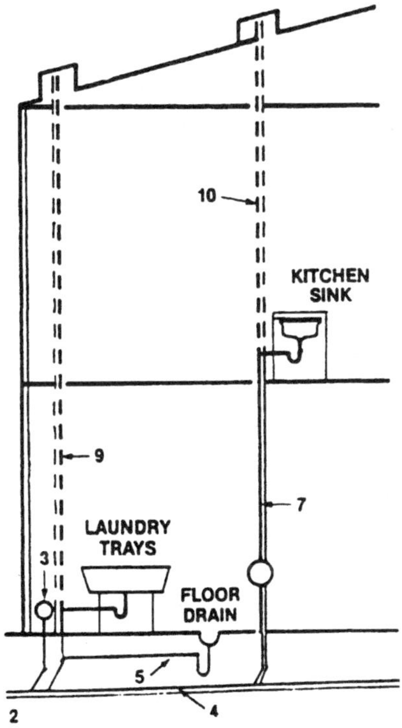

Figure 5-15 Two-story Plumbing Layout

If there is a change of use or occupancy, the codes require basic changes be made to meet the newer (and usually more restrictive) code regulations. This will be determined by the new construction classification. Egress requirements must also be looked at in relationship to the new remodeling or alteration. Such things as exterior exit requirements will be major factors in the framing design.

For example, sliding doors installed flush to the exterior wall are a common balance of design to meet code requirements and not detract from primary floor area. Multiresidential framing has further code requirements for exits and corridors, and mandates no projections into corridor space, as the required width of corridors must be unobstructed.

The exception to this is that handrails and doors, when fully opened, must not reduce the code-required egress width by more than 7 in. Doors in any position cannot reduce the required width by more than one-half and are framed with the same structural header assembly as shown in Fig. 5-16. Lintels for wider openings are shown in Fig. 5-17.

Openings for stair framing are shown in Fig. 5-18. Further details on stair framing are shown in Figs. 5-19 through 5-24. Other nonstructural framing projections such as trim and similar decorative features may project into the required width a maximum $1^1/_2$ in on each side.

Interior Wall Panels

Fiber-reinforced cement panels used in the finishing system allow for quicker occupancy and reduced labor and labor-related costs. Fiber-reinforced cement panels are tough, flexible, won't rot or buckle, and they're immune to water damage, snow, salt air, and termite attack. They are also the most fire-resistant siding money can buy.

Cement board installation, joint treatment, and base-coat application can be performed in one day. The exterior textured finish can be applied the next day to complete the system. Panelized interior wall construction of this type accommodates a variety of applications.

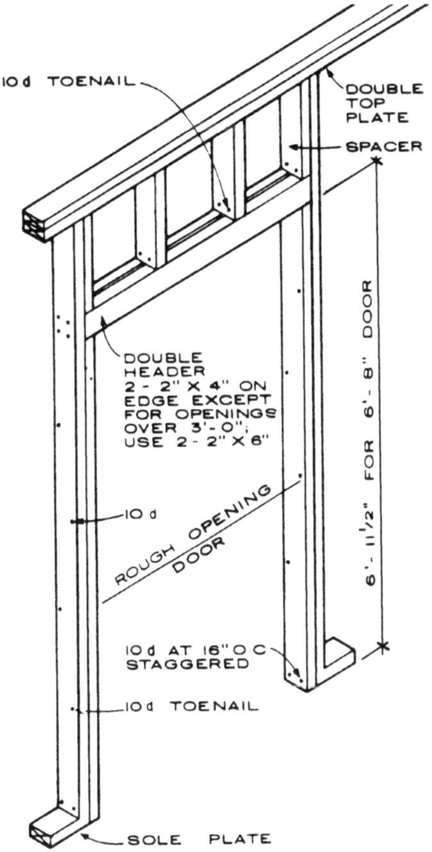

Figure 5-16 Door Opening

Structural heights can reach three stories, and clear spans can reach 48 ft. Square footage can range from 400 to 8500 sq ft. Panelized interior wall construction can also be employed for multistory and multiresidential construction and retrofit construction. Also, because the framework is doubled at each side of the 4 × 8 shear panels, the builder ends up with a stronger dwelling.

However, closed wood panels erected in a wall sometimes tend to grow due to swelling of the wood. In a 35-ft-long wall, swelling as much as $^1/_2$ to $^3/_4$ in in expansion from one

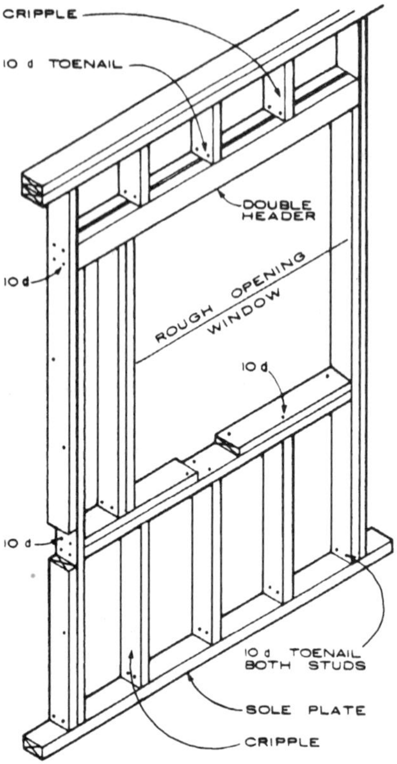

Figure 5-17 Window Opening

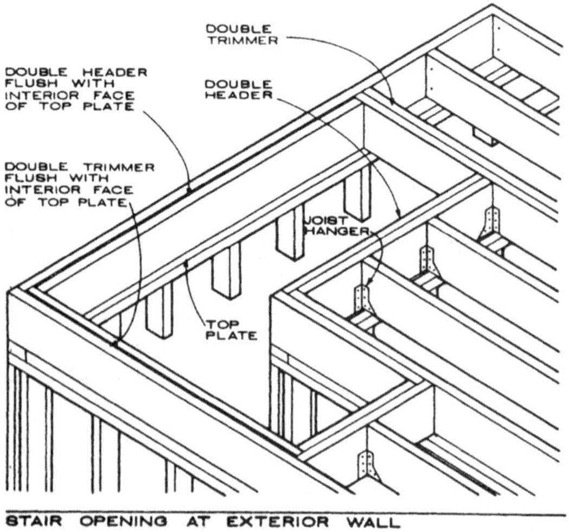

Figure 5-18 Stair Framing

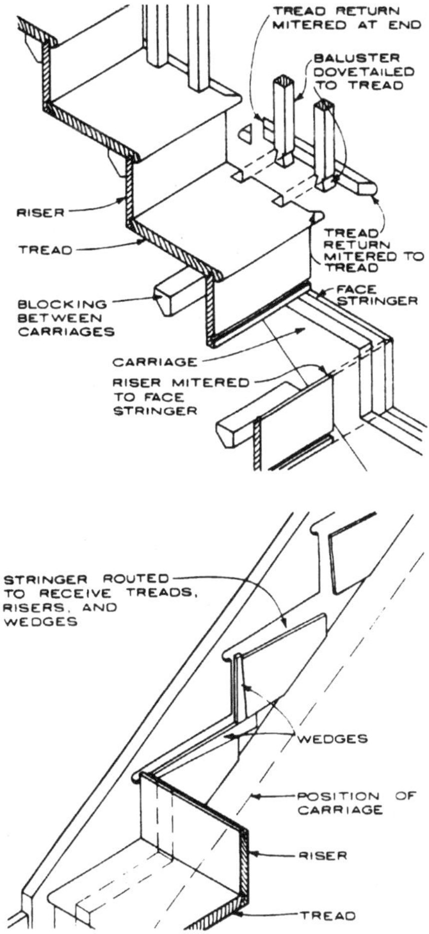

Figure 5-19 Stair Framing

end of the wall to the other can happen in areas of high temperature swings. This leads to significant problems with both the electrical and plumbing systems.

Bearing and Nonbearing Partitions Framing

Framing for interior bearing partition walls is shown in Fig. 5-25, and for nonbearing partitions in Fig. 5-26. A garage partition wall is shown in cross section in Fig. 5-27. Allowable header spans in wall openings are regulated by the building codes as shown in Tables 5-3 through 5-5.

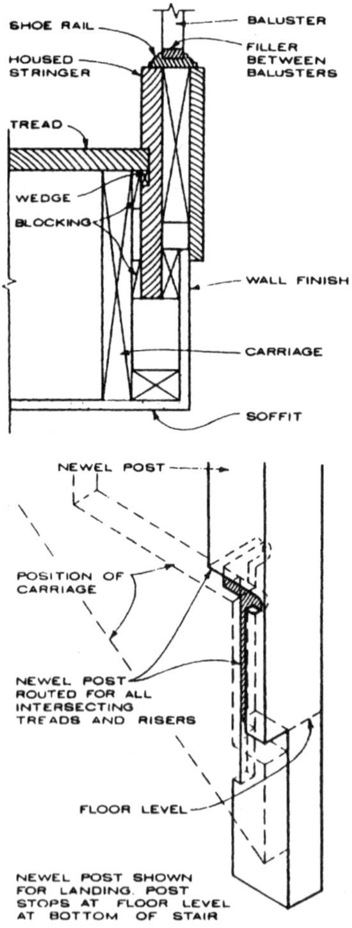

Figure 5-20 Stair Framing

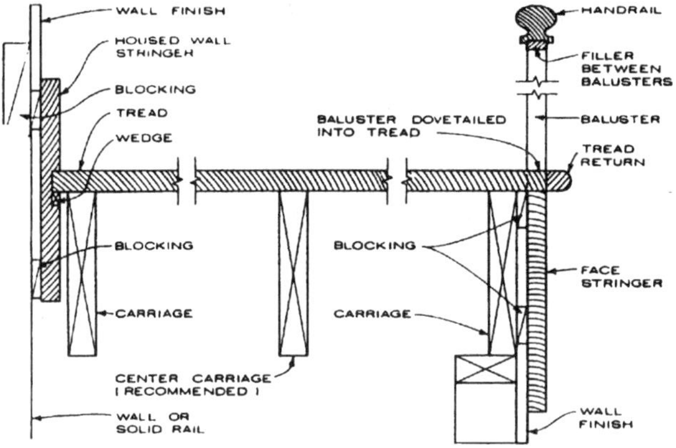

Figure 5-21 Stair Framing

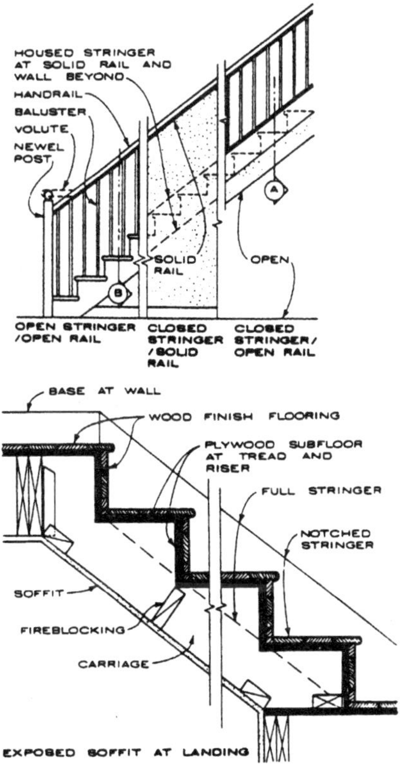

Figure 5-22 Stair Framing

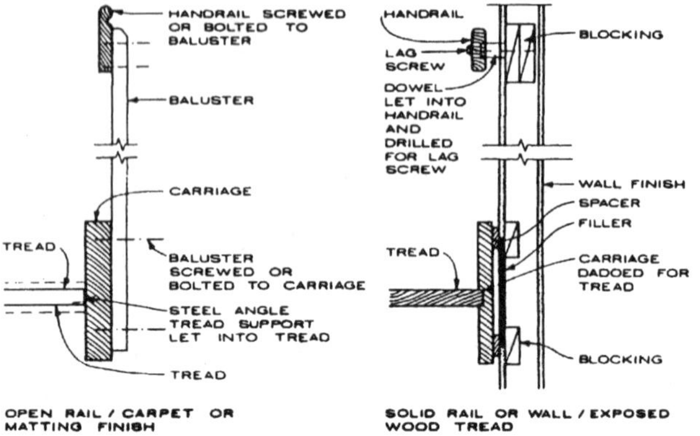

Figure 5-23 Stair Framing

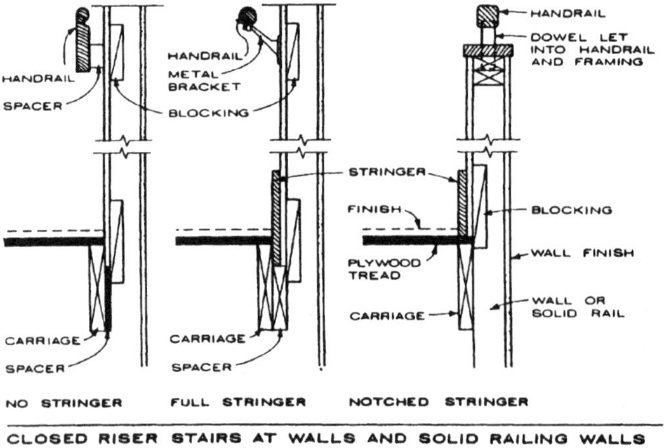

Figure 5-24 Stair Framing

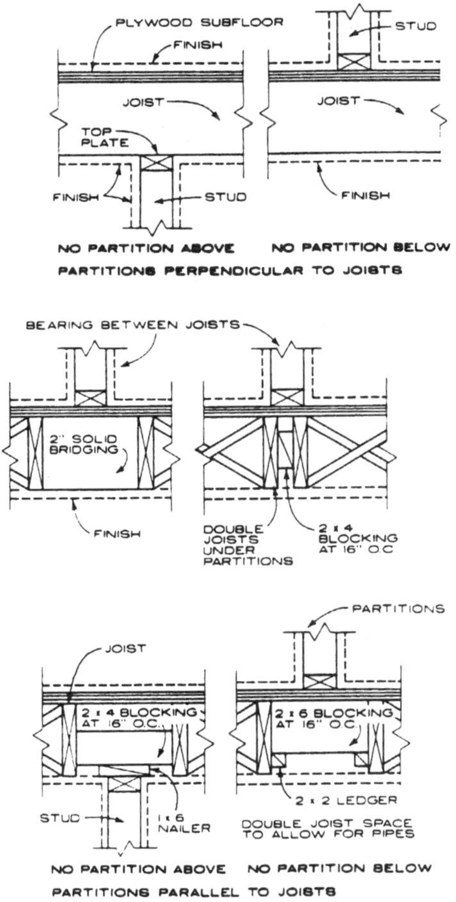

Figure 5-25 Bearing Interior Partition Walls

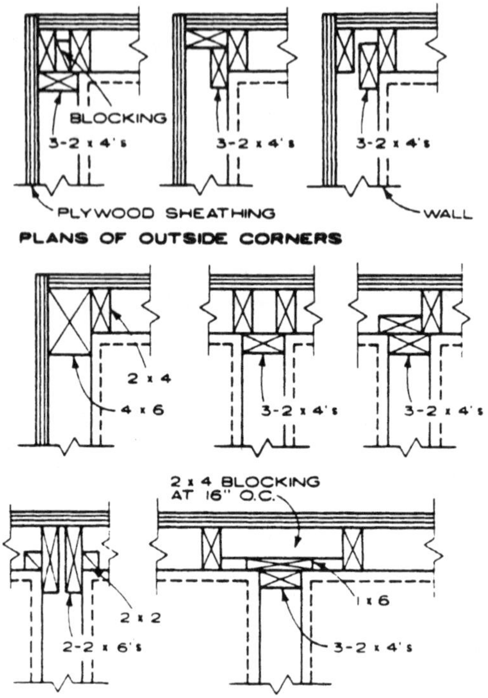

Figure 5-26 Nonbearing Interior Partition Walls

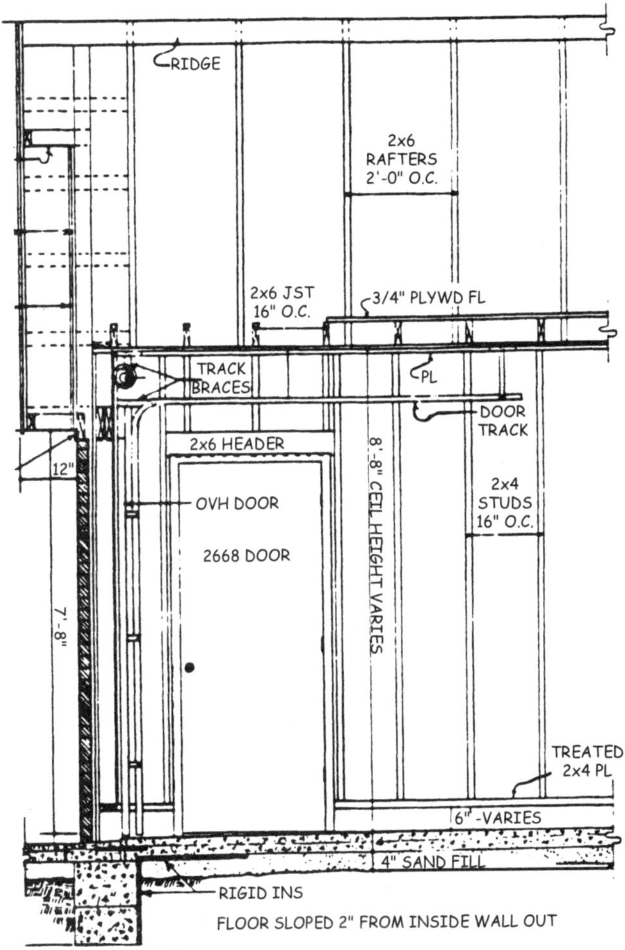

Figure 5-27 Garage Partition Wall

Table 5-3
Maximum Allowable Header Span

			Maximum Allowable Header Span (ft-in) Design Procedure: Single Top Plate Construction: Interior Bearing Walls								
Species, Grade	Header	Header	Nonstructural Sheath. House Depth (ft)			1/2" Insul. Bd. Sheath. House Depth (ft)			15/32" Plywood Sheath. House Depth (ft)		
Properties	Supporting	Size	24	28	32	24	28	32	24	28	32
Douglas Fir	One Story	2-2x4	3-5	3-2	3-0	3-10	3-7	3-4	4-7	4-4	4-2
Larch	Bearing	2-2x6	5-1	4-8	4-5	5-5	5-0	4-8	6-4	5-10	5-6
Southern Pine		2-2x8	6-7	6-1	5-8	6-10	6-4	5-11	7-8	7-1	6-7
No. 2		2-2x10	8-3	7-8	7-2	8-6	7-11	7-4	9-1	8-6	8-0
E = 1,600,00		2-2x12	10-0	9-3	8-8	10-1	9-6	8-10	10-5	10-0	9-5
F_b = 1200	Two Story	2-2x4	2-7	2-4	2-3	2-10	2-8	2-6	3-7	3-4	3-1
F_v = 90	Bearing	2-2x6	3-9	3-6	3-3	4-0	3-9	3-6	4-8	4-4	4-1
		2-2x8	4-10	4-6	4-2	5-1	4-8	4-5	5-8	5-3	4-11
		2-2x10	6-1	5-8	5-4	6-4	5-10	5-6	6-10	6-4	5-11
		2-2x12	7-5	6-10	6-5	7-7	7-0	6-7	8-1	7-6	7-0
	One Story	2-2x4	3-2	2-11	2-9	3-6	3-3	3-0	4-4	4-1	3-10
	Bearing	2-2x6	4-8	4-3	4-0	4-11	4-7	4-3	5-9	5-4	5-0
Hem Fir		2-2x8	6-0	5-6	5-2	6-3	5-9	5-5	7-0	6-5	6-0
No. 2		2-2x10	7-6	7-0	6-6	7-9	7-2	6-9	8-5	7-9	7-3
E = 1,400,00		2-2x12	9-1	8-5	7-11	9-4	8-8	8-1	10-0	9-3	8-7
F_b = 1000	Two Story	2-2x4	2-4	2-2	2-0	2-7	2-5	2-3	3-3	2-11	2-8
F_v = 75	Bearing	2-2x6	3-5	3-2	3-0	3-8	3-5	3-2	4-3	4-0	3-9
		2-2x8	4-5	4-1	3-10	4-8	4-3	4-0	5-2	4-9	4-6
		2-2x10	5-7	5-2	4-10	5-9	5-4	5-0	6-3	5-9	5-5
		2-2x12	6-9	6-3	5-10	6-11	6-5	6-0	7-5	6-10	6-5
Spruce Pine	One Story	2-2x4	2-11	2-9	2-7	3-3	3-0	2-10	4-1	3-10	3-7
Fir No. 2	Bearing	2-2x6	4-4	4-0	3-9	4-8	4-3	4-0	5-5	5-0	4-8
E = 1,300,00		2-2x8	5-7	5-2	4-10	5-10	5-5	5-1	6-6	6-0	5-8
F_b = 875		2-2x10	7-1	6-6	6-1	7-3	6-9	6-3	7-10	7-3	6-10
F_v = 70		2-2x12	8-6	7-11	7-5	8-9	8-1	7-7	9-4	8-7	8-1
	Two Story	2-2x4	2-2	2-0	1-11	2-5	2-3	2-1	3-1	2-10	2-6
	Bearing	2-2x6	3-3	3-0	2-9	3-5	3-2	3-0	4-0	3-9	3-6
		2-2x8	4-2	3-10	3-7	4-4	4-0	3-9	4-10	4-6	4-2
		2-2x10	5-3	4-10	4-6	5-5	5-0	4-8	5-10	5-5	5-1
		2-2x12	6-4	5-10	5-6	6-6	6-0	5-7	6-11	6-5	6-0

Table 5-4
Header Design Chart

Header Design Chart

(Douglass Fir-Larch, Southern Pine No. 2, Roof Load + 20LL + 10DL, Floor Load = 40LL + 10DL)

$F_b = 1200$ $E = 1,600,000$ $F_v = 90$

		Maximum Allowable Header Span (ft-in)								
Header Supporting	Header Size:	Nonstructural Sheath. House Depth (ft)			1/2" Insul. Bd. Sheath. House Depth (ft)			15/32" Plywood Sheath. House Depth (ft)		
		24	28	32	24	28	32	24	28	32
Roof	2-2x4	4-7	4-6	4-3	4-11	4-8	4-6	5-7	5-4	5-1
	2-2x6	6-8	6-4	5-11	6-11	6-7	6-4	7-7	7-3	7-0
	2-2x8	8-3	8-0	7-7	8-6	8-2	7-11	8-11	8-7	8-4
	2-2x10	9-10	9-6	9-3	10-0	9-8	9-4	10-4	10-0	9-9
	2-2x12	11-4	10-11	10-7	11-5	11-1	10-9	11-10	11-5	11-1
Roof + One Story (Bearing)	2-2x4	5-2	4-11	4-9	5-4	5-1	4-10	5-8	5-5	5-2
	2-2x6	6-9	6-5	6-0	6-11	6-7	6-3	7-4	7-0	6-8
	2-2x8	8-0	7-5	7-0	8-1	7-8	7-3	8-5	8-2	7-9
	2-2x10	9-3	8-9	8-3	9-4	9-0	8-5	9-7	9-3	8-11
	2-2x12	10-5	10-1	9-7	10-6	10-2	9-9	10-10	10-5	10-1
Roof + One Story (No Bearing)	2-2x4	4-8	4-5	4-3	4-9	4-6	4-3	5-1	4-10	4-8
	2-2x6	5-10	5-5	5-1	6-0	5-7	5-3	6-7	6-2	5-9
	2-2x8	6-9	6-3	5-11	7-0	6-6	6-1	7-6	7-0	6-7
	2-2x10	8-0	7-5	6-11	8-2	7-7	7-1	8-8	8-1	7-7
	2-2x12	9-3	8-7	8-1	9-5	8-9	8-3	9-11	9-3	8-8
Roof + Two Stories (Bearing)	2-2x4	4-8	4-5	4-3	4-9	4-6	4-3	5-1	4-10	4-8
	2-2x6	5-10	5-5	5-1	6-0	5-7	5-3	6-7	6-2	5-9
	2-2x8	6-9	6-3	5-11	7-0	6-6	6-1	7-6	7-0	6-7
	2-2x10	8-0	7-5	6-11	8-2	7-7	7-1	8-8	8-1	7-7
	2-2x12	9-3	8-7	8-1	9-5	8-9	8-3	9-11	9-3	8-8
Roof + Two Stories (No Bearing)	2-2x4	3-11	3-8	3-5	4-1	3-9	3-7	4-5	4-2	4-0
	2-2x6	4-8	4-4	4-0	4-10	4-5	4-2	5-3	4-11	4-7
	2-2x8	5-5	5-0	4-8	5-7	5-2	4-10	6-0	5-7	5-3
	2-2x10	6-4	5-11	5-6	6-6	6-0	5-8	6-11	6-5	6-0
	2-2x12	7-5	6-10	6-5	7-8	7-0	6-6	7-11	7-4	6-11

Table 5-5
Maximum Allowable Header Span

Maximum Allowable Header Span (ft-in)[1]
Design Procedure: Double Top Plate Construction: Interior Bearing Walls

Species, Grade Properties	Header Supporting	Header Size	Nonstructural Sheath. House Depth (ft)			1/2" Insul. Bd. Sheath.[2] House Depth (ft)			15/32" Plywood Sheath.[3] House Depth (ft)		
			24	28	32	24	28	32	24	28	32
Spruce Pine Fir No. 2 E = 1,300,000 F_b = 875 F_v = 70	One Story Bearing	2-2x4	3-3	3-0	2-10	3-7	3-4	3-1	4-4	4-0	3-9
		2-2x6	4-7	4-2	3-11	4-10	4-6	4-2	5-7	5-2	4-10
		2-2x8	5-9	5-4	5-0	6-0	5-7	5-2	6-8	6-2	5-9
		2-2x10	7-2	6-8	6-3	7-5	6-10	6-5	8-0	7-5	6-11
		2-2x12	8-8	8-0	7-6	8-10	8-2	7-8	9-5	8-9	8-2
	Two Story Bearing	2-2x4	2-5	2-3	2-1	2-8	2-5	2-3	3-3	3-0	2-9
		2-2x6	3-4	3-1	2-11	3-7	3-4	3-1	4-2	3-10	3-7
		2-2x8	4-3	3-11	3-8	4-5	4-1	3-10	4-11	4-7	4-3
		2-2x10	5-4	4-11	4-7	5-6	5-1	4-9	5-11	5-6	5-2
		2-2x12	6-5	5-11	5-7	6-7	6-1	5-8	7-0	6-6	6-1

1 in = 25.4 mm   1 ft = 0.305 m   1 psi = 6.8948 kPa

Notes

1. Linear interpolation for house widths not in table is permitted. For example, assume a 26 ft (7924 mm) wide house with 15/32 inch (11.9 mm) plywood sheathing - 2x6 header allowable header span = 7 ft - 5 in (2261 mm).
Header spans identified as having "no bearing" construction apply to both interior and exterior loadbearing walls which have tributary areas equal to one-half the house depth. Header spans identified with "bearing" construction apply only to exterior bearing walls with tributary areas equal to one-fourth of the house depth.
2. Sheathing or combined sheathing/siding having a minimum density of 18 pcf (288.3 kg/m^3)
3. Minimum 15/32 inch (11.9 mm) plywood sheathing or combined sheathing/siding applied between bottom of header, the top of the top plate and between lines of the broken vertical studs at the ends of the header and nailed to the header, top plates, cripples and studs - 6 inches (152 mm) o.c. at the edges and 12" (305 mm) o.c. at intermediate framing.

Notes

Notes

Chapter

6

Engineered Wood Components

Prefabricated Framing Assemblies

Prefabricated framing assemblies like roof trusses and floor trusses find widespread use in platform frame buildings because of their speed of erection, economy of material usage, and long spans. Most are light enough to be lifted and installed by two carpenters.

Manufactured wall framing panels have been adopted more slowly than roof and floor trusses, except by large builders who mass-market hundreds or thousands of houses per year. For the smaller builder, wall framing can be done on-site with the same amount of material as with panels and with little or no additional overall expenditure of labor, especially when the building requires walls of varying heights and shapes.

However, engineered wood structural panels are becoming more cost-effective each passing day in platform framing, so we will look at these in detail in the following section.

Engineered joists are an I-joist with structural wood flanges and structural plywood webs. Stiffeners at the bearings are installed in the field when required. Typically, the plywood is $^{3}/_{8}$-in Structural 1 placed in the joist in 4-ft lengths with the face grain running vertically. The plywood sections are butt jointed and glued to form a continuous web. The plywood flange connection is made by inserting the web into a groove in the center of the web face of the flange. All of the adhesives used should be waterproof exterior grade.

Structural Panels

Wood-based panel products have been utilized successfully as structural and decorative materials for over 60 years. Recent improvements in technology have expanded the types of panel products which can be labeled and used as *structural*. This term is used to mean that one of the intended functions of the panel product is to successfully resist the actions of externally applied forces on the panel for a specific application.

The American Plywood Association (APA) has set the performance standards for all engineered wood-based panels for framing sheathing and floor applications. These standards have been accepted by all of the major model building codes. The standards do not specify plywood but encompass any engineered wood-based panels that meet the code performance criteria. Performance standards of new engineered wood components continue to receive greater emphasis in the code revision committees.

This will simplify many structural design decisions, allowing selection between numerous alternatives on a common and more uniform basis. Several trade associations and manufacturers provide excellent up-to-date information on the characteristics and uses of engineered wood panel products. In particular, APA and the Engineered Wood Association offer a wealth of information and practices on the structural use of wood panels.

Structural panels are made from wood-based panel products that have been proved to be cost-effective in light wood framing applications with reliable long-term performance.

The new age of technology has also brought many improvements to the industry that have expanded the different types of panel products that are now specified as being of structural grade.

The term *structural grade* means that the panel product has been tested in an independent, sanctioned laboratory and can successfully resist the actions of externally applied forces on the panel for a specific application.

All wood-based structural panels are composite engineered wood components. They are a composite of wood elements like thin sheets of wood called veneer, wood particles, flakes, or fibers and an adhesive designed for either interior or exterior use. Plywood is the most commonly known structural product and is a glued assembly of cross-grain sheets of veneer, usually an odd number. Particleboard is a combination of wood particles and adhesive and is mainly used as floor underlayment and some specialty applications.

Waferboard is a variant of particleboard with flakes as the basic wood element. OSB (oriented strand board) is an engineered wood product that has wood strands, flakes, and fibers in layers oriented in different directions through the thickness of the panel. OSB competes very strongly with plywood in the residential and light wood frame commercial marketplaces.

A hybrid product, called composite plywood or composite panel, has face veneers with oriented strand or random particle-flake cores and is now available and meets code specs.

Many other types of panel products are available (often without added adhesive) that may have some structural capability such as hardboard, medium-density fiberboard, and insulation board. For example, wood fiber insulation boards contribute significantly to the racking strength of wall systems. However, these products are not generally intended to be structural in nature except as a by-product of their specific function in a building's structural framing system. No specific structural contribution is assumed.

Typical Applications

Engineered wood components are used to provide cost-effective stronger floors and straighter ceilings. Structural wood panels are typically used as sheathing to provide a strong, stiff covering over some relatively flexible and weak framework. In new residential and commercial construction, wood panels are most often used as roof sheathing and decking, floor decking and underlayment, wall sheathing, and exterior siding. Additionally, plywood is used in the manufacture of structural subsystems such as stressed-skin panels, built-up beams or columns, and other modular assemblies.

Because plywood has traditionally been the most used structural panel material, plywood performance has become the standard by which other panel products are judged. Many of the products described in the previous section can compete favorably with plywood both economically and in performance. Because of the many new products entering the marketplace, the designer should always be keenly aware of the advantages and disadvantages of alternative panels. Particular attention should be paid to three areas.

First, an accurate and complete specification of all the requirements of the material in terms of its necessary function and the environment in which it has to provide service must be developed. Second, accurate and reliable information (or assurance) of the expected properties and/or performance of the material must be obtained as input to a sound engineering design. Third, the material must be evaluated, not only in a strict engineering sense but also in terms of economics, ease of installation and maintenance, availability, and the long-term potential consequences of its failure.

Panel Standards

Because of the many choices the framer has with panel products, standards have been developed to ensure product quality and reliability. Standards for wood-based panel products are voluntary and have been developed by interested parties. However, building codes, local, and state governments often utilize standards in a legal sense to ensure an acceptable level of service and safety to the public. Consequently, both the architect and the framer should be aware of the potential use of these "mandatory" standards for a specific application.

There are two types of standards for wood-based panel products. The first is a commodity (or product) standard that describes what may be used in the manufacture of a panel product. It may state some minimum properties that the product must have to be acceptable, but these are usually only loosely tied to performance. The *Voluntary Product Standard for Construction and Industrial Plywood, PS1-83*, and the American National Standard *Mat-Formed Wood Particleboard*, ANSI A208.1-93, are examples of this type of standard. The second type of standard specifies a level of performance of a product in service, and in a specific application it generally does not attempt to regulate how the manufacturer should make the product.

For panel products, laboratory tests have been specified that have a direct relationship with the expected end use of the material. One test developed by the American Plywood Association for roof and floor panels simulates the actions of the heel of a construction worker or the load transmitted by a leg of a piece of furniture to the panels. Plywood product standard PS1-83 includes both types of standard concepts. Panels may be certified as meeting PS1-83 quality requirements either by an inspection of the manufacturing process or by periodic performance tests of random samples of finished panels.

Despite the influx of the new engineered wood panel products, plywood will continue to be the most commonly used panel product for structural use. It is such a perfect product for its application that the performance of newer panel products is judged against that of plywood. Consequently, this section further explains the various properties and uses of plywood in engineered wood components. As stated previously, plywood is a panel or sheet product composed of thin sheets of veneer, or plies. Plywood always has an odd number of layers, each layer consisting of one or more plies. These layers are glued together with the grain of adjacent layers placed at right angles to one another. The face layers are oriented with their grain parallel to the long dimension of the panel, with the crossbands thus having their grain parallel to the short sides of the panel.

All plywood is manufactured in sheets or panels in the modulus of construction standard of 4 × 8 ft (1.22 by 2.44 m). Other length sizes may be available for specific types of plywood, although widths other than 4 ft (1.22 m) are rare.

The tolerance for length and width dimensions of an engineered wood plywood panel is +0 and ×$^1/_{16}$ in (1.6 mm). This tolerance standard allows for some change in panel dimension with varying moisture conditions when gaps are left between panels in a sheathing application.

Plywood is available in a number of thicknesses from $^1/_4$ (6.4) to 1$^1/_8$ in (28.6 mm) and greater, with tolerance varying by size. For sanded panels, the tolerance is + $^1/_{64}$ in (0.40 mm) for thicknesses of $^3/_4$ in (19 mm) or less or + 3 percent for $^3/_4$ in (19 mm) thick.

For unsanded, touch-sanded, and overlaid panels, the tolerance is + $^1/_{32}$ in (0.8 mm) for thicknesses of $^{13}/_{16}$ in (20.6 mm) or less or + 5 percent for larger thicknesses.

Many variables in the composition of plywood and its use in engineered wood panels affect its properties and performance as a structural panel. These include the number and thickness of the layers and plies, the species of veneer (they may not all be the same), the grain orientation of the plies, the quality or grade of veneer, and the type of adhesive used in binding the plies.

Combinations of these variables enable the manufacturer to tailor the product for specific uses. Hence many different types of plywood and plywood panels are available. The framer must understand the composition of various types of plywood to select the appropriate panel for a specific application.

Veneer and Adhesive Panels

The thin sheets of wood that are laminated to form plywood are usually rotary-peeled. That is, a long knife, acting somewhat like a potato peeler, peels the veneer from a spinning log. The veneer is then clipped to size and rapidly dried to a low moisture content. Some repairs may be made to upgrade the quality of the sheet. With one type of repair called *plugging*, a defect such as a knot or pitch pocket is cut out and patch fit into the remaining hole. Smaller sheets of veneer may be joined or spliced together to form the necessary sizes. Veneer to be used in core stock may be strung or stitched together with nylon string or thread.

The sheet of veneer is then visually graded. The grade of the veneer is an important factor in determining the grade of the plywood panel. The veneer grades are constructed so that a grade C veneer may be upgraded to grade B. Hence similarities in terms of strength and stiffness exist between grades A and C and also between grades B and D veneer. Grade C-plugged veneer is an upgrade of C that does not meet all the requirements of grade A. The type 4 adhesive used to laminate the veneers influences the durability of the resulting panel. The most common adhesives are phenolic resins, which are synthetic adhesives that cure under high temperature and must be hot-pressed. They are waterproof and will not be degraded by microorganisms. Phenolic resins may be extended or thinned by mixing them with water and other ingredients to produce a lower-cost adhesive that has some water resistance but is not waterproof.

Solid wood, and particularly veneer, has different strength and stiffness properties along the grain than across it. A sheet of veneer stressed in tension may be over 100 times stronger parallel to grain than perpendicular to grain. The modulus of elasticity across the grain is approximately one-thirty-fifth of that parallel to grain.

By cross-laminating, plywood can be made more homogeneous in terms of strength properties than a sheet of solid wood of the same thickness. However, there will still be a strong and a weak orientation to a panel. This is a result of the plies, with grain parallel to the stress having greater resistance than those perpendicular to the stress.

Because there is a greater thickness of veneer placed parallel to the face grain, the strong direction is parallel to the face grain of the panel with the weak orientation 90° from that. Additionally, the cross-sectional properties are different in the two orientations.

Wood is a hygroscopic material and will change dimensions when the surrounding air changes relative humidity and temperature. It is very anisotropic in its shrinking and swelling characteristics, and dimensional changes across the grain are many times greater than parallel to grain. Because of plywood's cross-banded nature, plywood in-plane swelling is much less than that of solid wood and is usually less than 2 percent when a dry panel is saturated. Swelling through the thickness is similar to that of solid wood (2 to 5 percent dry to green) and varies with species.

For normal construction, swelling problems can be virtually eliminated by providing a $^1/_8$-in (3-mm) gap between panel edges and ends. These values should be doubled if the moisture content is expected to be high in service. In modern panel construction, it is known that the use of panel clips, which are used to minimize deflection differences between adjacent panels, will help in ensuring adequate spacing.

Plywood panels also have many of the advantages and properties of solid wood, such as light weight [35 lb/ft^3 (640 kg/m^3)], ease of working, high strength-to-weight ratio, and ease in fastening. Additionally, plywood is split-resistant and has excellent shock absorbency characteristics. Like wood, plywood is an insulator and has low thermal conductivity when dry.

It is also an electrical insulator (dielectric) when dry and has excellent chemical resistance. (All of these properties vary by species and moisture content.) It may be overlaid or surfaced with a resin-impregnated fiber [high-density overlay (HDO) or medium-density overlay (MDO)] to yield a hard, smooth surface for painting or for greater resistance to chemicals and abrasion. Panels may be overlaid with other products including metals to provide surfaces for specific applications or finishes.

One of the major advantages of plywood panels is their fastening characteristics. Because of the crossbands, nails may be placed near the edge of the panel without danger of panel splitting. This is one of the keys to its success as a sheathing material. The total effectiveness of bolts, screws, staples, and other fasteners is also enhanced by the crossbanding.

Plywood Panel Classification

Plywood is manufactured in two broad types based on resistance to moisture and weathering. *Exterior* plywood has a 100 percent waterproof glue and has a restriction that no veneer grade below C is allowed. Exterior-type panels should be used when they will be exposed to weather or to moisture. *Nonexterior* plywood is for applications that are not permanently exposed to weather. There are three common classifications of nonexterior panels.

Exposure 1 panels have a waterproof bond and are made for conditions where long construction periods may result in an unprotected exposure or where high moisture conditions may be encountered in service. APA-rated sheathing, Exposure 1 is probably the most common grade-marked structural panel in the market, but it is not designated for permanently exterior application. Under PS1-83, these panels may be identified as "Interior plywood with Exterior glue."

Exposure 2 panels (also called "Interior plywood with Intermediate glue" under PS1-83) are designed to have some moisture resistance but may not have a fully waterproof adhesive or the necessary veneer quality to fully resist more than a brief delay before protection, even if painted.

Interior-only panels lack any information about adhesive water resistance in their trademark and should be used only for protected applications. The only recommended allowable stresses in panels are keyed to the type of plywood and the adhesive used.

Plywood Panel Grades

Plywood is manufactured with different face, back, and core veneer qualities for many different applications. Grades of construction and industrial plywood are usually identified by the face and back veneer grades (e.g., C-D, A-B, A-C) or by a descriptive name (e.g., Marine, Underlayment, Structural 1). Panels may be sanded for better appearance, touch-sanded to make panel thickness more uniform, or left unsanded. The appearance and specialty grades include sanded panels and siding that do have structural properties but are manufactured specifically for appearance rather than strength and stiffness. Appearance panels are generally designated by the panel thickness, the face and back veneer grades, and some indication of wood veneer species.

Other grades (sometimes called *engineered grades* or *structural-use panels*) of panels are designed specifically for strength and stiffness and include APA-rated sheathing and APA-rated Sturd-I-Floor. The former includes C-D Exterior or C-C Exterior and also may be called Structural 1 or 2. (Structural 2 is rarely manufactured.) The latter is a combination subfloor-underlayment. These products may be conventional plywood, composite plywood, or OSB (oriented strand board).

These panels are generally unsanded or touch-sanded and are designated by thickness and a number indicating allowable spans between supports for floor and roof applications. The species is generally not specified, but some species restrictions may be inherent in the grade as specified by PSI. For example, Structural 1 is limited to group 1 species.

Grademarks

Plywood that is produced in conformance with PS1-83 may be grade-stamped by a qualified inspection and testing agency. This agency certifies that the quality of the panel conforms to the product standard, a proprietary specification, such as APA's performance ratings, or both. A program of quality control, inspection, and testing by the independent agency ensures the user of getting plywood of the grade and quality specified.

Several independent agencies provide this service, but the majority of plywood manufactured in the United States is grademarked by the APA. Figs. 6-1 and 6-2 show typical APA grademarks. Fig. 6-3 shows grademarks for treated lumber.

Section Properties

In many sheathing applications, the architect or building designer does not need to know the section properties or allowable stresses for plywood. Building code regulations and trade association information are often organized to require a minimum amount of calculation. However, for those "nonstandard" applications and for a clear understanding of the numbers in design tables, effective section properties must be used.

Because plywood is essentially an orthotropic plate, it has different properties in its three principal directions. These properties are directly related to thickness and number of veneers, the adhesive, and the species used in the panel. Plywood manufacturers have some latitude in the number and thickness of veneers used to make a panel of a certain thickness in conformance with PS1-83. Since the potential geometries are too numerous for practical tabulation, "effective" section properties have been computed for a range of panel composition.

Species

The term *softwood plywood* is often considered synonymous with *structural plywood*. However, over 70 different species,

Figure 6-1 Grademark

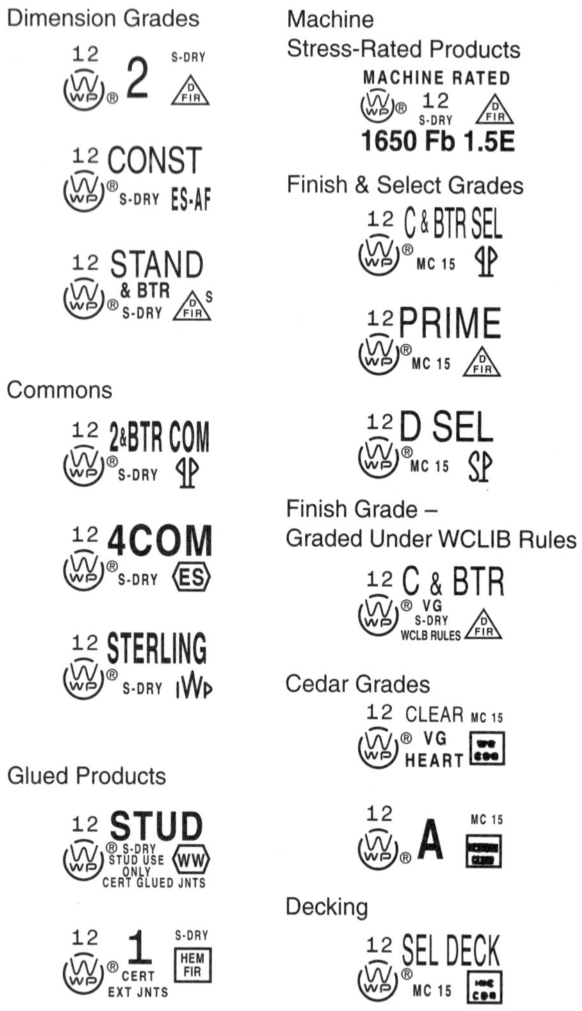

Figure 6-2 Grademarks

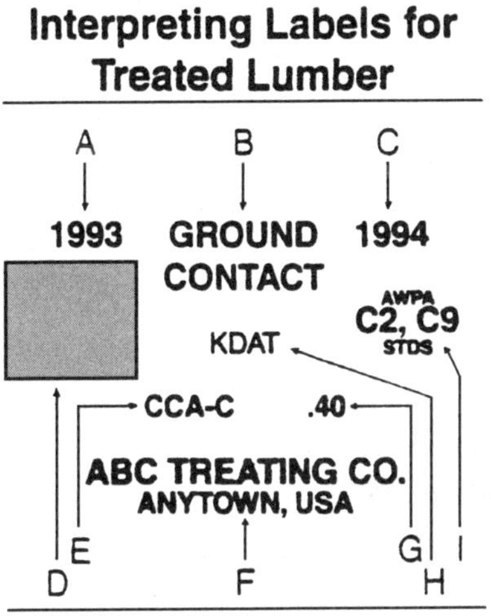

A) Year of treatment
B) Proper exposure conditions
C) Year of treatment
D) Trademark of inspection agency approved by the American Lumber Standards Committee (ALSC)
E) Preservative used for treatment
F) Treating company and location
G) Retention level
H) Dry or KDAT, if applicable
I) Applicable American Wood Preserver's Association (AWPA) Standard

Figure 6-3 Treated Lumber Grademark

including several hardwoods, may be used in construction and industrial plywood conforming to PS1-83. The product standard divides these species into five groups based on strength and stiffness with group 1 as the strongest. Plywood of groups 1 through 4 has been assigned allowable stresses for use in structural applications.

Because various species may be mixed in the veneers of a given panel, the group number in the grademark is based on the species of the face and back veneers. In setting allowable stresses the inner plies are assumed to be group 4 regardless of their actual composition.

Span Ratings

Plywood in engineered wood panels may be used in applications where forces are applied either perpendicular to the panel or in the plane of the panel. The former are typically sheathing applications covering two or more supports such as siding, roofing, or flooring. Because of these common applications, it is possible to rate plywood on the basis of its strength and stiffness parallel to face grain.

The span rating is two numbers separated by a slash and is found in the grade trademark stamp on plywood manufactured for sheathing purposes. The first number is the allowable span in inches for roof sheathing with face grain across supports and with the panel continuous over two or more spans. The second number is the allowable span for plywood when it is used as floor sheathing with the same conditions.

The higher the index number, the greater are the strength and stiffness. This index is designed to eliminate the need for structural calculations to select the proper panel for a sheathing application. The actual maximum spans allowed may be varied by the local building code authority.

For APA-rated Sturd-I-Floor, single-floor applications (combined subfloor and underlayment), only one span rating is given for the panel application. This number represents the maximum support spacing for combination subfloor and underlayment under an average residential load. PS1-83 prescribes veneer thickness, species groups, quality, etc., for each span rating. There is also an alternative provision to qualify a plywood panel construction for a span rating on the basis of performance tests. Hence it may not be possible to determine the composition of a panel with a particular span rating. However, performance tests ensure that all panels with the same rating may be freely substituted regardless of internal construction. PS1-83 applies only to plywood panels. Span ratings are also assigned to composite panels through performance testing prescribed in PS2-92 or APA-PRP 108.

In many applications plywood may be subject to loads both normal to the panel and in the plane of the panel. In the latter case, the panel thickness is critical. Also, on many jobs it may be desirable to maintain a constant thickness for plywood with a certain span rating. Consequently, both the span rating and the thickness must be compatible, and both should be specified.

Mechanical Fastenings

Many different types of mechanical fasteners have been used successfully with plywood and other similar panel products. For most applications, nails, staples, wood screws, or bolts are sufficient. In sheathing applications, a fastening schedule (Table 6-1) is usually specified rather than using individual joint ultimate strengths. These fastening schedules are based on test results and commonly successful construction practice. They will typically specify the type of fastener to be used and the maximum spacing between fasteners. Most building codes and regulatory agencies specify sheathing fastening to be used in specific applications.

The strength of mechanically fastened plywood joints depends on the type of plywood and the substrate to which it is to be fastened.

Design values and other specifications for some pneumatic or mechanically driven staples and nails are given in National Evaluation Report, NER-272. This report contains fastening specifications for structural panels used in walls, floors, and roofs, including shear walls and diaphragms. Other specific design values may be found in the *National Design Specifications For Wood Construction* or local building codes.

Recommended or required fastening schedules should always take precedence over Table 6-1 for specific applications such as the design of structural diaphragms or shearwalls. For conventional light frame construction, nails and screws should be spaced 6 in (152 mm) apart at panel edges. Panel fasteners should be a minimum of $^3/_8$ in (9.5 mm) from panel edges.

H-shaped panel clips (ply clips) are commonly used for supporting panel edges in roof construction. They may be

Table 6-1

Fastening Schedule

Connection	Fastener	Number or Spacing
Joist to band joist, face nail	16d common	3
Joist to sill or girder, toe nail	8d common	3
Bridging to joist, toe nail each end	8d common	2
Ledger strip	16d common	3 at each joist
1x6 or less subfloor to each joist, face nail	8d common	2
Over 1x6 subfloor to each joist, face nail	8d common	3
2 inch subfloor to joist or girder, blind and face nail	16d common	2
Sole plate to joist or blocking, face nail	16d common	16" o.c.
Top or sole plate to stud, end nail	16d common	2
Stud to sole plate, toe nail	8d common	4
Doubled studs, face nail	10d common	24" o.c.
Doubled top plates, face nail	10d common	16" o.c.
Top plates, lap and intersections face nail	–	2-16d or 3-10d common
Continuous header, two pieces	16d common	16" o.c. along each edge
Ceiling joists to plate, toe nail	8d common	3
Continuous header to stud, toe nail	8d common	3
Ceiling joists, laps over partitions, face nail	–	3-16d or 4-10d common
Ceiling joists to parallel rafters, face nail	–	3-16d or 4-10d common
Rafter to plate, toe nail	8d common	3
1-inch brace to each stud and plate, face nail	8d common	2
1x8 or less sheathing to each bearing, face nail	8d common	2
Over 1x8 sheathing to each bearing, face nail	8d common	3
Built-up corner studs	16d common	24" o.c.
Built-up girders and beams, of three members	20d common	32" o.c. at top and bottom and staggered 2 ends and at each splice

Connection	Fastener	Number or Spacing
Hardboard Lap Siding Over Sheathing	10d corrosion resistant with minimum shank diameter of 0.099 inch and minimum head diameter of 0.240 inch	16 inch o.c. at top and bottom edges
Hardboard Panel Siding Direct to Studs	6d corrosion-resistant with minimum shank diameter of 0.092 inch and minimum head diameter of 0.225 inch	6 inch o.c. at edges and 12 inch o.c. at intermediate supports
Hardboard Panel Siding Over Sheathing	8d corrosion-resistant with minimum shank diameter of 0.092 inch and minimum head diameter of 0.225 inch	6 inch o.c. at edges and 12 inch o.c. at intermediate supports

used in place of lumber blocking or tongue-and-groove edges. When required, one clip is generally placed between each support except for 48-in (1220-mm) spans, where two are required to function as effective edge support.

Composite Structural Panels

Particleboard panel products consist of wood particles in the form of flakes, strands, wafers, chips, shavings, fibers, etc., that are combined with synthetic resins or some other bonding system. In mat-formed particleboard, the mixture of wood particles and binders is laid down in a mat that is hot-pressed to increase the density of the mat and cure the adhesive.

The particles may be oriented to some degree to provide some directionality of strength properties along and across the panel. Oriented particles may be laid down in orthogonal layers similar to plywood veneer. Different types of particles may be used in the same panel to improve surface properties and create a more layered particle product.

Standards

Particleboard panel products may be manufactured under a voluntary standard, ANSI A208.1-1979, *Mat-Formed Wood Particleboard*. This standard covers the physical and mechanical property requirements for grades of mat-formed board. The only structural particleboard produced under this standard is manufactured house decking. The National Particle Board Association should be contacted for recommended use of this product.

Composite panel products may be manufactured to meet the American Plywood Association's performance standard PRP-108 or the U.S. Voluntary Product Standard PS2-92. If the panel meets the requirements for structural adequacy, dimensional stability, and bond durability, it may qualify as APA performance-rated structural-use panel as sheathing, Structural 1 sheathing, or single floor. This rating indicates that the panel is sufficient for the specified application and not that the different products all have the same properties.

Waferboard

Waferboard is an exterior bonded structural panel product that has largely been replaced by oriented strand board but may still be available in some areas. It is made of wood flakes whose length in direction of the grain is at least $1^1/_4$ in (31.8 mm) and is bonded with a phenolic adhesive. Wafers are usually randomly oriented to make an in-plane isotropic panel. However, they may be deliberately aligned to increase the strength and stiffness properties parallel to alignment. Those perpendicular to alignment are reduced in this case.

Allowable properties for waferboard have been established by the Structural Board Association (Canadian agency). These properties and associated product applications have been approved for use in the United States through a National Evaluation Service report assigned to the Structural Board Association.

Oriented Strand Board (OSB)

OSB is a panel product that is comprised of strandlike particles of wood that are directionally oriented. Three to five layers of these strands are arranged in a panel at right angles to one another. This makes the panel more homogeneous in its in-plane properties and improves dimensional stability and fastening characteristics. OSB panels may be rated under the APA performance specifications for sheathing applications. Oriented strand material also may be used as corestock in composite plywood.

Composite Panels

This panel product has been commercially available for several years and is a composite of veneer and wood-based materials. Typically, the grain direction of the face veneer is in the long direction of the panel. The grain of the reconstituted wood core may be oriented at 90° to the face veneer and could be one of several types of reconstituted wood.

Oriented strand cores are common. These products are marketed under a variety of names, but Comply, an APA proprietary trade name, is common. These products are usually rated under the APA performance specifications or PS2-92.

Structural Sheathing Applications

By their very nature, panel products are efficient in providing both surface coverage and strength requirements in frame construction. This section discusses typical applications of panel products as sheathing-type load-carrying elements. Loads are assumed to act normal to the panel surface as opposed to diaphragms or shearwall-type loadings.

Historically, the most common wood panel product for floor, wall, and roof sheathing has been plywood. Other composite panels can be used, since the function of the panel is the same in the sheathing system regardless of its composition. In recent years, composite panels and OSB have become very common in the structural-use panel market. This is the primary reason for the advent of performance standards for panel products.

Plywood and panels with directional characteristics are normally used in the strong direction and are assumed to be continuous over two or more spans. One exception is in stressed-skin panel systems, where the weak direction is often employed. These recommendations conform to most building codes but may differ with requirements of specific localities.

Flooring Systems

Panel products may have several different uses in a flooring system. As a subfloor in a two- or more-layer floor system, the sheathing is connected directly to the joists or beams that provide the major system support. In this system, an underlayment is also used on top of the subfloor to provide a smooth surface for floor coverings and for additional strength and stiffness.

Wood-strip flooring, lightweight concrete, or other coverings with some structural capacity may preclude the need for underlayment. A combined subfloor and underlayment such as APA's Sturd-I-Floor is a single-layer product that serves as both subfloor and underlayment.

Performance requirements for floor sheathing are usually established by the building code, lending institution, or government agencies. Minimum requirements are generally specified for deflection between joists or beams under a uniform or concentrated load. Additionally, the sheathing should safely support all design uniform loads and any specified concentrated loads. There are differences in requirements among agencies.

Subfloor

The sheathing grades of plywood are frequently used as subflooring. Tongue-and-groove (T&G) edges provide interlocking a flush subfloor surface and self-spacing T&G also simplify installation. These include PS1-83 plywood grades C-D Exposure 1 and C-C Exterior. APA-rated sheathing (Exposure 1 or 2) is designed specifically for this application and may include performance-rated composite panels. Vertical notches in the panel's tongue and ends allow water to drain during construction. Nominal sizes range from $^{19}/_{32}$ to $1^1/_8$ in.

The span rating is a guide to the use of a panel in a specific application. If lightweight or gypsum concrete is to be used over the subfloor, then panels marked $^{42}/_{20}$ may be used when joists are spaced 24 in o.c. (610 mm). A $^{32}/_{16}$ index is the minimum that should be used for 16-in (406-mm) spacing. All recommended spans assume the panel to be continuous over two or more spans with grain of face veneer parallel to the span.

The free edges of the subfloor should be tongue-and-groove (T&G) or supported by blocking unless a separate underlayment layer is installed with joints offset from those of the subfloor. Lightweight concrete or wood-strip flooring also may preclude edge support.

Many code agencies accept particleboard and waterboards for use as subfloors or combined subfloor and underlayment. This acceptance will vary by locality. Typical recommendations are a maximum span of 16 in (406 mm) for $^5/_8$-in (15.9-mm) and $^{21}/_{32}$-in (16.7-mm) panels and a 19.2-in (488-mm) span for $^3/_4$-in (19.0-mm) panels. The edges must be T&G or otherwise supported.

Underlayment

This layer provides a smooth surface for nonstructural floor coverings and increases floor stiffness. It also must have sufficient hardness to resist indentation during normal use. Either plywood or grade PBU particleboard is typically used. Underlayment-grade plywood has C-plugged face veneer and spliced inner plies to resist indentation.

Plywood underlayment is laid down in the same direction as the plywood subfloor so that the grain of the face veneer is perpendicular to the joists.

However, the end joints and side joints should be staggered with respect to the subfloor panels. Particleboard is quite commonly used for underlayment and is typically applied at right angles to a plywood subfloor. The required thickness of underlayment may be specified by code or by the type of surface it covers.

Underlayment should be installed immediately before the finish flooring to minimize damage during construction. It also should be water-resistant if used in a high-moisture-content area such as a bathroom.

Combined Subfloor and Underlayment

It may be economical to use a panel product that will serve as both subfloor and underlayment. Labor savings are possible, but good continuity between panels is necessary and either 2-in (51-mm) wood blocking or tongue-and-groove edges are required at panel edges between supports.

Engineered wood underlayment has a smooth, durable surface that is an excellent base for thin, resilient floor covering. It retains dimensional stability under high moisture conditions better than plywood, OSB, and glulam underlayment. In addition, engineered wood underlayment is easy to install—standard modulus 4 × 4 ft panels have the fastening pattern printed on the face.

The American Plywood Association (APA) has a proprietary product called "Sturd-I-Floor" that is grademarked with a span rating. This rating assumes a minimum uniform live load of 100 lb/ft^2 (4.79 kPa) and a $^1/_{360}$ span deflection limit for a maximum span less than or equal to 32 in (813 mm). Composite panels are acceptable under Sturd-I-Floor grademark. A special panel product, Sturd-I-Floor 48 o.c. (formerly called 2-4-1), is recommended for spans of 32 or 48 in (813 mm or 1.22 m). It is $1^1/_8$ in (28.6 mm) thick and should be used with 4-in nominal width joists only if the span is 48 in (1.22 m). The allowable uniform live load in this case is 55 lb/ft^2 (2.63 kPa). If the spans are 32 in (813 mm), nominal 2-in-wide framing may be used, and the allowable load will increase.

In recent years, OSB has been accepted by all major code authorities in the United States and Canada for some applications. This product has some excellent characteristics if used properly and is accepted in some areas as combined subfloor-underlayment, subflooring, or underlayment. Local building codes should be checked for its acceptability in a specific area. Typically, a $^5/_8$-in (15.9-mm) or thicker panel is required for 16-in (406-mm) spacing and a $^3/_4$-in (19.0-mm) or thicker panel for 19.2-in (488-mm) spans. Nailing requirements are similar to those for plywood.

Glued Floor Systems

To enhance the performance of the floor system, the sheathing layer may be field-glued to the joists or beams. This enables the sheathing and joists to function as a T beam, resulting in a significant increase in floor stiffness and, potentially, load capacity. Glued floors result in increased stiffness and fewer squeaks, nail popping, and costly callbacks for repairs.

Both single- and multiple-layer floors may be glued, but panels with tongue-and-groove edges are most favored. The adhesives should meet performance specifications such as APA's AFG-01 and be easily installed in the field. The contractor must see that OSB panels with sealed surfaces and edges are compatible with the glue.

These floor systems start with engineered wood joists that are designed to start straighter and stay straighter than solid sawn joists, and resist shrinking, twisting, splitting, warping, and crowning.

Heavy-Duty Plywood Floors

PS1-83 plywood can be used to support loads that are heavier than those assumed in the span-rating system. Performance-rated panels may be used only if they are marked PS1-83. For example, in a light frame commercial, if a light-duty storage area is to be built over a series of offices and the floor of this area will function as the office ceiling and will be exposed with no finish covering, a single-layer floor can be used over 2 × 8 joists spaced 16 in (406 mm) o.c. with the face grain perpendicular to the supports. If the local building code authority mandates a design live load of 125 lb/ft^2 (5.98 kPa), what PS1-83 plywood or performance-rated panel could be used?

APA-rated Sturd-I-Floor is designed for combined subfloor-underlayment application. A 16-in (406-mm) joist spacing and greater than 100 lb/ft^2 (4.79 kPa) design live load, a 20 on-center span-rated Sturd-I-Floor panel, or APA-rated sheathing with a $^{32}/_{16}$ index will suffice. If a finish floor is to be installed on these panels, then a thin underlayment panel should be added.

If PSI plywood is specified, then APA-rated sheathing with a span index of $^{32}/_{16}$ or 16 on-center Sturd-I-Floor can be used instead. Note that by specifying a PSI trademark, you are requiring an all-veneer panel product.

Wall Sheathing Applications

Plywood is commonly used in wall construction to provide structural rigidity or to provide both rigidity and aesthetic characteristics. in the first case, plywood is used as an underlayment on which siding, brickwork, or other materials are attached. It is generally used at corners to provide racking resistance but may cover the entire structure. In this mode it distributes the wind load to the wall framing members. Spans and thicknesses are shown in Tables 6-2 and 6-3.

APA-rated sheathing is appropriate, as are the PS1-83 plywood grades of C-D Exposure 1 or C-C Exterior. The face grain may be applied in either direction. Shearwall action will require blocking at panel edges.

Plywood is also available as a single-layer or surface panel which may have a variety of textures or patterns in the face and yet perform as a structural element. APA 303 siding may be applied directly to studs or to a nonstructural panel sublayer and will provide the strength and rack resistance necessary to eliminate the cost of diagonal bracing or a structured sublayer. Medium-density overlay (MDO) also provides this

Table 6-2

Allowable Spans for Wood Structural Panel Wall Sheathing

Panel Span Rating	Panel thickness (in) and Construction	Maximum Stud Spacing (in) with Exterior Covering Nailed to	
		Stud	Sheathing
12/0, 16/0, 20/0 or Wall-16oc	5/16	16	16
24/0, 24/16, 32/16 or Wall-24oc	3/8, 7/16,15/32 1/2	24	24

Table 6-3

Minimum Thickness of Wall Sheathing		
Sheathing Type	Minimum thickness (inches)	Maximum stud spacing (inches)
Wood boards	5/8	24
Fiberboard	7/16	16
Wood Structural Panel	According to Table 2308.1B	
2-M-1 or 2-M-W Particleboard	According to Table 2308.1C	
Gypsum sheathing	1/2	16
Gypsum wallboard	1/2	24
Reinforced cement mortar	1	24

1 in = 25.4 mm

advantage as well as that of a smooth, paintable surface. Local code authorities should be consulted for details.

Glulams

Plywood may be combined with lumber or glued-laminated timbers (glulam) to form I- or box-shaped beams. Glulams are available in thicker, wider, and longer dimensions than solid sawn lumber. They also offer more load-carrying capacity and a higher degree of design freedom. By combining exterior adhesives and ultrasonically graded veneers, glulams create a strong engineered beam. With the shortage of large dimensional lumber, glulams are an affordable and available alternative for the modern wood framer.

They are typically available in standard thicknesses of $1^1/_2$ and $3^1/_2$ in and depths of $7^1/_4$ to 18 in. These nailed and glued flexural members have a very high strength-to-weight ratio and are commonly used for 20- to 80-ft (6.10- to 24.4-m) spans. They may be straight or cambered and consist of one or more vertical plywood webs joined to dry lumber flanges. The lumber flanges are designed to carry most of the bending moment, and the plywood is designed to carry the shear stresses. Stiffeners are used between flanges to prevent web buckling and to transfer bearing forces.

The plywood and lumber are joined with adhesives using clamps, presses, nails, or other means to obtain necessary pressure for proper bonding. Only positive mechanical pressure (not nails) can be used in lumber flanges of more than one lamination. End joints in flanges and webs may require scarf or finger joints for splicing.

Flange and Web Design Properties

Axial tension or compression allowable stresses (not bending allowable stresses) for the lumber flange are determined for the specific grade of lumber or for the glulam combination to be employed. These are modified for applicable durations of load. If two or more pieces of lumber are to be laminated together, the relative location of the joints may affect the design stresses. Full design stresses may be used for properly made scarf or finger joints as outlined in the *Plywood Design Specification (PDS)*.

Butt joints may be used in the flanges, but some reduction in the effective cross-sectional area is necessary. If the butt joints in the two adjoining laminations are spaced 50 times the lamination thickness t apart, the full cross-sectional area of one lamination may be used. Other spacings will require some reduction as specified in the PDS.

The APA specification on the fabrication of plywood beams requires that butt joints be spaced $30t$ apart in adjoining laminations and $10t$ apart in nonadjoining ones. This results in an effective cross-sectional area of 90 percent of one layer for a two-lamination flange or 90 percent of two layers for a three-lamination flange. The allowable stress in tension for butt-jointed members in multiple laminated flanges is 80 percent of that for the grade. This is in addition to any reduction in effective area.

Web Properties

The plywood webs will contribute to the beam moment of inertia a thickness equal to that of the layers of veneer with grain parallel to the beam length. Scarf joints of $^1/_8$ slope will transmit 100 percent of axial or bending stresses. A slope of $\frac{1}{5}$ (which is the limit) or flatter transmits 75 percent of tension or bending stress and 100 percent of compression stress.

Butt joints backed by a glued plywood splice plate with its grain perpendicular to the joint may be used. The plate should be equivalent in thickness and quality to the web.

No reduction in panel shear strength is necessary if the glued plywood splice plate is at least 12 times as long as the panel thickness.

Design Procedures Checklist, Plywood

The following steps are necessary in the design of a lumber-plywood beam. The designer usually knows the load, span, deflection criteria, environmental conditions, and any architectural or manufacturing considerations such as beam depth or appearance restrictions and material availability:

- □ Given the above, determine a trial size using approximate methods.
- □ Determine the bending moment and the required section properties.
- □ From horizontal shear, determine plywood thickness.
- □ From the shear at the flange-web joint, choose adequate dimension and geometry to meet specs.
- □ Determine section necessary to meet deflection requirements. Also determine necessary camber if required.
- □ Determine size and spacing of bearing and intermediate stiffeners.
- □ Determine the details of any necessary splices.
- □ Determine requirements for lateral supports.

Glued Stressed-Skin Panels

Plywood and lumber also can be combined to form a flat, shallow assembly designed to resist flexural stress from loads applied normal to the panel surface. These stressed-skin panels are not unique to wood frame construction and have been used in aircraft, ships, and vehicles as well as buildings. They are a particularly efficient form of construction when material cost or weight is a consideration. They are usually prefabricated in a factory and are manufactured with a precision and quality that minimizes field installation time and trouble.

Stresses-skin panels may be flat or curved and are often used in folded-plate or curved-roof structures as well as in wall and floor systems. They may be either hollow web components or a sandwich with a lightweight core. Polystyrene or polyurethane foams or paper honeycombs have been used in the latter case. In curved panels, plywood may be used as a core material as well as for the skins.

The stressed-skin plywood-lumber panel gains its efficiency from having the plywood skin take most of the bending stresses while also performing as sheathing. The lumber webs space the skins, thereby increasing the moment of inertia; they also carry the shear stresses.

Panels may be one- or two-sided or have lumber T-flanges. In any event, the lumber members or stringers are rigidly connected to the flanges so that the panel functions as a series of built-up beams. Elastomeric adhesives have been used in panels subject to impact loads or stress concentration.

However, there is no simple way of designing panels with elastic connections, and their use is somewhat limited to trial-and-error applications. The design methods described in this section assume that the flanges and webs are rigidly connected.

Fabrication

The size of an individual panel is governed by either the application or the sizes of available plywood and lumber. The lumber stringers could be multiple-member laminates or scarf-jointed, but generally this is not economical or practical to accomplish. For specific cases, this generality may not hold, but usually the length of the panel is limited to the available length of the lumber.

The plywood skins may be spliced or scarfed to provide the necessary surface area. Scarf joints of 1 in 8 or flatter transmit full allowable stress. A slope of 1 in 5 transmits 75 percent of the allowable stress, and steeper joints are not recommended. Finger joints may be used if supported by test data. While PS1-83 plywood is currently recommended for use in stressed-skin panels, successful designs with structural composite panels will inevitably evolve.

Some bridging or blocking is generally required in stressed-skin panels. Blocking will support splice plates for joining skins, distribute concentrated loads, and align stringers during construction. Headers provide the same function at the ends of the panel. Blocking may be notched to provide ventilation. If panels are used in a pitched roof with stringers perpendicular to the support rafters, blocking must be used to improve racking resistance. One-sided panels may require bridging for lateral stability purposes.

The plywood on the bottom of a two-sided panel may be on a variety of decorative forms for aesthetics. If appearance is a factor, $^{5}/_{16}$ in (7.94 mm) should be the minimum thickness to avoid a wavy appearance. Usually the plywood is applied with face grain parallel to the stringers, which negates use of span ratings.

The lumber web members and the plywood or lumber flanges are rigidly connected with casein, phenol, resorcinol, or melamine adhesives. Only kiln-dried lumber is used. The choice of adhesive is particularly important when wet-use in-service conditions are expected, and only waterproof adhesives should be used in these applications. Either interior or exterior plywood grade may be used depending on the moisture condition in service.

Lumber-plywood panel assemblies are often slightly unsymmetrical in cross section to provide an effective tongue-and-groove arrangement for mating with other panels. The tongue, or male stringer, may be increased in width to provide rolling shear area and a nailing surface when mated to an adjacent panel. The groove, or female stringer, is recessed in from the edge of the skins and may be reduced in thickness, since the stringer in the mating panel will carry some of the load.

Design Considerations

The properties of the lumber stringer should be those recommended by Section 5 of the *Plywood Design Specifications* and may be modified for load duration and moisture content. Lumber stringers are normally surfaced for gluing and should be considered $^1/_{16}$ in (1.59 mm) smaller in depth (perpendicular to gluing surface) for one-sided panels and $^1/_8$ in (3.18 mm) smaller for two-sided and T-flange panels.

Stressed-skin panels are generally assumed to be in flexure with the span parallel to the stringer. For a rigidly glued composite of plywood and lumber with differing elastic properties, a transformed section must be computed. That is, the effective width of the lumber web is "transformed" on the basis of the ratio of the stringer MOE (modulus of elasticity) to the stringer skin MOE. Skins of differing stiffnesses also should be transformed to a reference stiffness. This gross moment of inertia I based on the transformed section assumes that the full width of the panel is effective. This section property is used for deflection and shear calculations.

In computing the bending stress, only the cross section of material with grain parallel to the span is considered effective. The area covered by an unspliced butt joint is ineffective, as is any area not covered by a splice plate (i.e., that area immediately over the stringer). Plywood in stressed-skin panels is usually oriented with face grain parallel to stringers.

Plywood that is rigidly connected to lumber stringers may have an effective width for bending stress purposes that is less than the actual width. This is because thin plywood may dish toward the neutral axis if the stringers are widely spaced. Limitations are placed on the effective width in accordance with a basic spacing b in the code table.

If a clear distance between stringers is less than the *b* in the code tables, no reduction is necessary. If it is exceeded, the distance between stringers that is considered effective is limited. The maximum clear distance between stringers is limited to twice the basic spacing. Note that the section properties may be different for deflection and bending stress calculations.

For computation of the rolling shear stress, it is necessary to compute a statical moment of the area outside the critical shear plane. Since rolling shear is a cross-grained shear, the critical plane lies in the cross-grained layer nearest the neutral axis. The critical statical moment Q for the top skin is usually the largest.

Allowable Stresses and Deflections

For stressed-skin panels, the published values of plywood and lumber elastic moduli should be increased 10 and 3 percent, respectively. This increase removes the allowance for shear deformation inherent in the published values. Shear deformation is considered separately in stressed-skin panels.

The lumber values published in the *National Design Specification for Wood Construction* should be used with appropriate modifications. Repetitive member bending values may be used if stringer spacing is 24 in (610 mm) or less, and there are at least three members. The modulus of rigidity G may be taken as 0.06 of the true stringer bending MOE.

Design Procedures Checklist, Panels

Lumber-plywood stressed-skin panels are designed on a trial-and-error basis. Some recommendations for typical panel configurations are available from the APA or from fabricators. The following checklist should be used for analysis or design for a specific application. It is assumed that the designer knows the load, span, deflection criteria, environmental conditions, regulatory guidelines, component properties, and availability of materials.

- □ After selecting a trial size, compute the section properties for deflection computations.
- □ Determine allowable load based on deflection including that of the top skin.
- □ Compute section properties for bending stress.
- □ Determine allowable load based on bending stress including the necessary skin or lumber splices.
- □ Determine the statical moment for rolling shear, and determine the allowable load based on the allowable rolling shear stress.
- □ Determine the statical moment for lumber horizontal shear and allowable load based on this property.
- □ If the panel is to be used in a wall, buckling or compression should be checked as well as combined bending and axial load.

Specific designs may require a check of factors other than those outlined above. If all materials in the panel have similar moduli of elasticity, an approximate design method may be used. However, the final trial sizes should be checked with a more thorough analysis.

Curved Panels

Stressed-skin panels also may be curved to function either as flexural members or as arch segments. The difference between the two classifications is the development of horizontal thrust at the supports in the arch segment. Arch panels are usually constructed with spaced ribs and may be either a fixed-fixed arch or a two- or three-hinged arch.

The APA recommends panels of medium curvature ($^3/_1$ to $^8/_1$ span-to-rise ratios) as the most practical. There are limits on the bending radii of plywood panels when they are dry. Some fractures may occur at this recommended minimum with typical plywood. Tighter radii may be achieved with wetting or steaming.

The design of spaced-rib arch and flexural curved panels is very similar to that of flat stressed-skin panels.

Other Stressed-Skin-Type Panel Products

The stressed-skin concept utilizes strong, relatively thin sheets designed to carry the tension and compression stresses due to bending. These skins are separated by a core designed to space the skins and carry the shear. Lumber cores in the form of stringers have been described in the preceding sections. However, other materials such as foams or paper honeycombs are used as well. These products are often called sandwich panels.

In the design of flat or curved sandwich panels, it is assumed that the core does not contribute to the bending stiffness of the panel. The section properties are thus determined assuming that only the skins are effective. These types of panels are often used in walls or other systems where combined bending and compression conditions exist. Hence, global and local buckling effects must be checked as well as bending and shear stresses.

Radial stresses in curved sandwich panels should be kept below one-third of the allowable shear stress of the core. Bearing stresses may also be important in sandwich panels, and points of contact with concentrated loads or reactions should be carefully designed.

Multiple layers of plywood may be glued together to form a solid plywood curved panel. The interior plywood may be glued over the full area of the skins or only in strips to reduce fabrication costs.

Applications

Glued plywood structural components have many advantages during and after construction. High strength-to-weight ratio, efficient use of materials, uniform size, production-line quality, and reduced overall construction time can be claimed by these products.

Stressed-skin panels are commonly used in roof and floor systems over 12- to 32-ft (3.66- to 9.75-m) spans. Two-sided panels may often be premanufactured with insulation installed, and the bottom skin may be a ready-to-finish ceiling. One-sided panels offer the advantage of allowing easy installation of a mechanical service system.

Folded-plate structures utilize either nailed or glued plates in a shell system that relies on its shape to support the horizontal and vertical forces. Those structures can be used in spans up to 100 ft (30.5 m) or more. Panel thickness is governed, by rafter spacing and shear adjacent to the rafter supports. The panels are prefabricated and are readily assembled in the field using either common or specially designed connectors and bearing hardware. The horizontal thrust of the plates is taken at the end walls and/or with horizontal tie rods.

Folded plates also may be designed to radiate out from an unsupported apex to supporting exterior walls or other supports near the building perimeter. Another form of construction that uses this type of plate is a space frame. Space frames usually incorporate a combination of plates that interact to support vertical loads without beams or trusses. The webs or skin of the shell structure transmit shear to the boundaries by diaphragm action.

Engineered wood rim board is a strong, cost-effective choice for engineered floor systems. Designed to match engineered floor joists, engineered wood rim board is available in $1^1/_8$ in nominal thicknesses and depths of $9^1/_2$ and $11^7/_8$ in, and lengths of 14 and 16 ft. The 16-ft length makes it quick and easy to install.

Diaphragms and Shearwalls Limitations

All structures are subjected to lateral forces generated by the wind. Buildings in specific geographic locations may additionally be subjected to lateral forces generated by the movement of their foundations during an earthquake. Other lateral forces such as those caused by earth pressure or blast loads also may act on structural frames. These design concepts are applicable in designing the frame resistance system of a building with wooden elements for any lateral force, and minimal emphasis will be given to the specific source of the lateral force.

The magnitude of the lateral force caused by either wind or earthquake that must be considered in a design as acting on the frame as a whole or on the frame's components is defined by the applicable building code. A large volume of technical literature is available to assist the design engineer in applying the code to specific design situations.

The applicable forces will simply be given in the examples used, since it is assumed that the designer either already has the necessary skill to determine the forces for a specific design situation or will refer to the appropriate literature for the needed assistance to develop this skill.

Wall, floor, and roof subsystems sheathed with wood products such as boards or plywood can be designed to function within the total building system to resist lateral forces in addition to supporting vertical loads. These subsystems, as detailed in a previous section in this book, are referred to as shearwalls (vertical elements) and diaphragms (horizontal elements).

A critical consideration in the proper functioning of diaphragms and shearwalls is the connections needed to ensure that these elements act as integral units and that these units in turn are interconnected with the building as a whole to ensure the proper functioning of the entire structure.

The detailed design of wood framing components and their connections is the subject of other chapters of this handbook. This section will consequently focus on the methods for determining the magnitude of forces on the structural panel elements of diaphragms and shearwalls and on the connections between elements. Once these forces are known, the design of the elements and connection proceeds according to the wood engineering principles discussed in other chapters.

Diaphragm Types

The earliest designed diaphragms utilized board sheathing laid diagonally. These diaphragms were regarded analytically as analogous to a truss with the boards acting as web members. Board sheathing is no longer used extensively in wood-frame construction. Panel products, particularly plywood, have largely replaced individual boards for roof, subfloor, and wall sheathing. See Table 6-3. The change has been beneficial in an engineering sense to wood-frame construction because a panel product will inherently provide stronger and stiffer diaphragms than those made of boards and can be engineered for high capacity.

Plywood sheathing panels have directional characteristics due to the grain orientation of the individual veneers that have been glued together to make up the panel. It is customary, since plywood is stronger and stiffer parallel to the panel's long dimension, to install plywood sheathing with the panel's long axis perpendicular to the framing members.

There are exceptions to this practice, such as panelized roofs. It is also customary to offset the end joints in adjacent rows of panels. The end joints are to occur over framing members. Plywood sheathing panels are typically 8 ft long, and to minimize cutting and waste, this requires a spacing of framing members, which cycles with 8 ft. Common framing member spacings are thus 12 in (305 mm) (eight per panel), 16 in (406 mm) (six per panel), and 24 in (610 mm) (four per panel). Other spacings include 48 in (1220 mm) (two per panel), 32 in (813 mm) (three per panel), and 19.2 in (488 mm) (five per panel).

Testing of plywood-sheathed assemblies as diaphragms has shown that panel thickness, panel layout, and the relationship to framing member orientation, framing member size, fastener size, and spacing all influence the shear carrying capacity of a diaphragm. These findings have been incorporated into tables developed by the American Plywood Association and adopted by building codes.

Other panel-sheathing product manufacturers have followed the lead of the plywood industry in testing their products in diaphragm applications and have obtained building code recognition of the diaphragm shear capacity of their products.

The long edges of the panels pose two separate design considerations. The first involves loads applied normal to the panel surface. A load on one panel, particularly a concentrated load near the panel edge, can cause the loaded panel to deflect relative to the adjacent panels. This problem is customarily overcome by providing a tongue-and-groove edge joint on panels used for floor sheathing. Because of the relative cost of a tongue-and-groove joint and the common use of thinner panels for roof sheathing, tongue-and-groove plywood is not typically used in roof construction.

Roof panel edges requiring support typically have this support provided by using ply clips or supporting wood members called *edge blocking*. Supporting wood members also can be provided for panels used in floors that lack a tongue-and-groove joint. The need to take some design action for loads applied normal to the panel, along the panel edges for certain combinations of panel thickness and member spacing, is covered by requirements in building codes that panel edges be "blocked." Panels that do not need edge support can be left unblocked.

The second design consideration involves the in-plane loads between panels functioning as part of a diaphragm. In-plane loading parallel to a joint between panel edges causes one panel to displace relative to the adjacent panel. This displacement can be minimized by attaching the edge of one panel to the edge of the adjacent panel in some manner. Various techniques have been employed, the most common being to place a secondary wood member below the joint between panels and to nail each panel edge into this common member.

Diaphragm tests have established that assemblies with panel edges interconnected to each other in this manner have greatly enhanced in-plane shear capacities. Diaphragms are thus referred to as *blocked* or *unblocked* depending on the presence or absence of attachment between panels for in-plane shear transfer.

The use of the same terminology, blocked and unblocked, for both in-plane and normal load considerations has caused confusion on occasion. It is important to keep in mind that the loadings are different with the two types of blocking, and the design solutions may be correspondingly different.

Since most panel products will swell as the moisture content of the panel increases, it is recommended that all panels be laid with gaps at their ends and edges. For plywood, a spacing of $^1/_8$ in (3.18 mm) is recommended for all edges, and other panel products may require different gaps. The producer's literature should be consulted.

Since panels are manufactured with tolerances that do not allow the panel to exceed the 8-ft (2.44-mm) and 4-ft (1.22-m) dimensions, some allowance is indirectly provided for spacing between panels. In long or wide structures it may be necessary to cut an occasional panel in order to maintain the recommended spacing between panels and to keep their edges properly centered over framing and blocking members.

Restrictions on Engineered Diaphragm Use

The depth-to-span ratios of wood diaphragms are limited. These limits are usually applied not only to the diaphragm as a whole but also to individual segments of a diaphragm, such as, for example, the solid area adjacent to openings. The depth is taken in the direction of the load, and the span is the distance between supports providing the diaphragm reactions.

Wood diaphragms are restricted in their application in buildings where an eccentricity between the center of the applied load and the resisting element(s) causes the diaphragm to want to rotate in a plane about a vertical axis. The usual situation involves a building with an all-glass facade and solid walls on each side and at the back. The building code restrictions are as follows:

"In buildings of wood-frame construction where rotation is provided for, the depth of the diaphragm normal to the open side shall not exceed 25 ft (7.62 m) nor two-thirds the diaphragm width, whichever is the smaller depth. Straight sheathing shall not be permitted to resist shears in diaphragms acting in rotation.

Exceptions:

1. One-story wood-framed structures with the depth normal to the open side not greater than 25 ft (7.62 m) may have a depth equal to the width.
2. Where calculations show that diaphragm deflections can be tolerated, the depth normal to the open end may be increased to a depth-to-width ratio not greater than 1.5/1 for diagonal sheathing or 2/1 for special diagonal-sheathed or plywood diaphragms. In masonry or concrete buildings, lumber and plywood diaphragms shall not be considered as transmitting lateral forces by rotation."

Subdiaphragms

The subdiaphragm concept was introduced into design practice following the 1972 San Fernando earthquake in California, which demonstrated weaknesses in the design approaches being used to anchor masonry and concrete walls. The concept awaits the test of experience, and this experience may prove that some combining of main and subdiaphragm design forces is necessary.

Loads on a diaphragm frequently are imposed on the diaphragm by walls along the diaphragm's edge. These loads may come from wind pressures on the walls or from the inertial forces of the walls, which are generated by the response of the building to the earth movement in an earthquake. Special consideration should be given to how these localized forces are anchored and distributed into the sheathing of a diaphragm.

Reliance should not be placed for anchorage on the connectors attaching the sheathing to the framing members. These attachments may already be working at or near capacity in carrying the diaphragm shear, and imposition of the localized wall load on these same connectors may result in a premature failure of the diaphragm.

Current practice for the anchorage of masonry or concrete walls requires such walls to be anchored to the diaphragm for a minimum of 200 lb/ft (2.92 kN/m) or for the computed component force, whichever is larger. If the anchorages are more than 4 ft apart, the compatibility of the wall and the diaphragm between anchorages is to be investigated. Continuous ties are also required between the diaphragm chords to distribute the localized forces into the diaphragm.

In many buildings it is not feasible to have continuous ties every 4 ft across the width of a building. The solution to such a design problem is to select a portion of the main diaphragm adjacent to the wall being anchored and to design that portion as a special diaphragm to perform the anchorage function. Such a selected portion of the main diaphragm is referred to as a *subdiaphragm*. The span-to-depth limitations applied to diaphragms are also applied to subdiaphragms. The subdiaphragms then span between the continuous ties between chords.

The use of a subdiaphragm to anchor a concrete wall is a principle that can be used to anchor any type of concentrated load. The design intent is to anchor the particular building component adequately to prevent a localized failure. Since the unit wind or earthquake design loading on a component is normally larger than the design loading on the entire structure, it is assumed that the anchorage problem can be addressed independently of the whole building design.

This permits the subdiaphragm to be designed for the anchorage loads alone, and the subdiaphragm can then be designed as a part of the main diaphragm without considering possible additive features of the two loadings. The most severe design conditions considered independently would govern the connection design within the subdiaphragm.

The subdiaphragm concept should not be regarded as applicable only to loads applied at the edges of a diaphragm. It is equally applicable to other types of problems.

Restrictions on Engineered Shearwall

Shearwalls are restricted in the ratio of their height to in-plane length. These ratios are generally interpreted as applying to both single and multistory buildings using the total wall height for multistory buildings. Unsuccessful efforts have been made to develop an engineering consensus on how to apply these ratios to multistory buildings in which the shearwalls do not align vertically between floors and/or the wall lengths vary between floors. It is recommended that the wall length for a wall on a given story, regardless of how it aligns with other shearwalls, be at least as long as the limiting ratios would require if the walls did align.

The rationale for these limiting ratios is that they represent engineering judgments as to the length of shearwall required to limit story drift under lateral loading to a reasonable amount. Wood-sheathed shearwalls are considered to be relatively flexible. This initially led to confining their use in bracing buildings with concrete or masonry walls to buildings on only one story. This restriction has subsequently been modified to permit wood shearwalls to be used to brace two-story buildings with masonry or concrete walls provided the following restrictions are observed:

1. Wall heights cannot exceed 12 ft (3.66 in).
2. Horizontal diaphragms are not considered to transmit lateral forces by rotation or cantilever action.
3. Deflections of horizontal and vertical diaphragms do not permit per-story deflections of supported masonry or concrete walls to exceed 0.005 times each story height.
4. Structural panel sheathing in all horizontal diaphragms has all unsupported edges blocked. Plywood sheathing for both stories of vertical shearwalls has all unsupported edges blocked and for the lower story walls has a minimum thickness of $1/2$ in (12.7 mm).
5. There are no out-of-plane horizontal offsets between the first and second stories of panel shearwalls.

Notes

Notes

Chapter

7

Roof Construction

Wood Light Frame Roofs

Many different types of roof designs are used in modern construction (see Fig. 7-1), however, most house roofs are either flat or peaked. The most common design is the *gable roof*, consisting of two planes that meet at the central peak ridge beam and slope down the building's longest exterior walls. The triangular sections at the ends are the gables that give the structure its name. Dormers are often added to gabled roofs to add ventilation and light.

A *hip roof* is similar to the gable roof except that all four sides are sloped. A *flat roof* generally overhangs the exterior walls of the house. On many houses, the various types of roofs are combined and variations on the basic styles are used.

Most roofs contain structural elements similar to those in a gable roof. The primary members are the rafters. The tops of the rafters are cut diagonally to fit flush against the ridge beam that runs the length of the roof at its peak. The ridge beam sits in a vertical ridge beam saddle attached to the center of the front and rear exterior walls.

The bottoms of the rafters are notched with two cuts, a seat cut and a plumb cut, to fit flush on the top plate doubler.

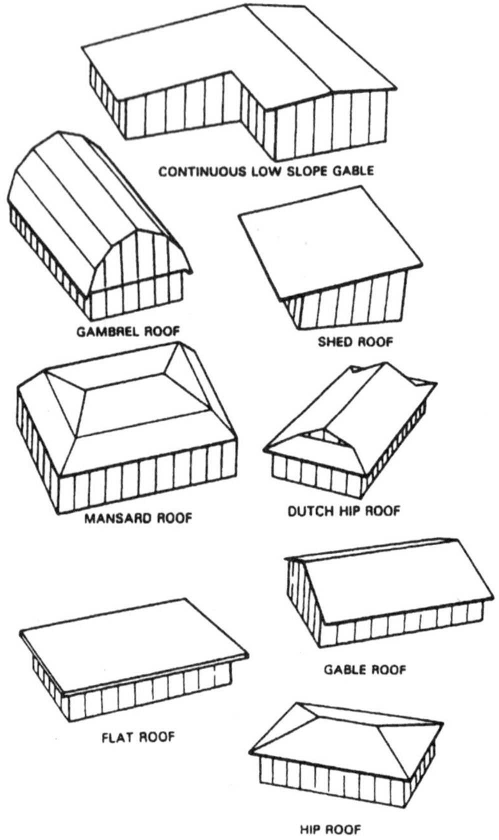

Figure 7-1 Roof Types

Below the ridge, collar ties that are made of construction grade 1 × 6s or 1 × 8s span the space between opposing rafters to keep them from spreading.

Roof sheathing, consisting of plywood panels or of tongue-and-groove boards, is laid over the rafters and a layer of 15- or 30-lb asphalt-impregnated felt is laid over the sheathing. Shingles or other roofing materials cover the felt.

With wood shingles or shakes, both of which need ventilation, rows of narrowly spaced furring strips may be nailed to the rafters over or in place of solid sheathing. At all joints, metal flashing must be added to make the roof watertight.

Flat roofs, by contrast, have no separate rafters; instead, the ceiling joists of the top story support the roofing. The joists are usually wider than normal, often 2 × 10s or even 2 × 12s. If the roof overhangs the exterior walls of the house on the sides parallel to the run of the joists, *outlook rafters* are used to create the overhangs.

They must be tied to a doubled ceiling joist, and the distance from the doubled joist to the outside wall is usually twice the overhang. In a flat-roofed post-and-beam house, heavy beams—or doubled or tripled reinforced boards—take the place of the ceiling joists.

Stick-Built Roof Framing

Stick-built roofs are built from structural framing rafters (Figs. 7-2 through 7-4). The standard types of roofs in western platform framing are often combined to make roofs suited to the covering of complex floor plans and building shapes. Gable and hip roofs are often combined to produce this architectural look.

For structural stability, gable and hip roofs must be securely tied by well-nailed ceiling joists to make what is in effect a series of triangular trusses. If the architect wants to eliminate the ceiling joists to expose the sloping underside of the roof as the finished ceiling surface (vaulted ceiling), a beam or bearing wall must be inserted under the ridge unless a system of exposed ties is designed to replace the ceiling joists.

While the college-graduate architect or engineer would find it difficult to lay out the rafters for a pitched roof using trigonometry, the framing carpenter, without resorting to mathematics, has little problem making the layout if the *pitch* is specified as a ratio of *rise* to *run*. Rise is the vertical dimension and run is the horizontal. In the United States, pitch is usually given on the architect's drawings as inches of rise per foot (12 in) of run. An old-time lead framing carpenter uses these two figures on the two sides of a framing square to lay out the rafter. See Figs. 7-5 through 7-7.

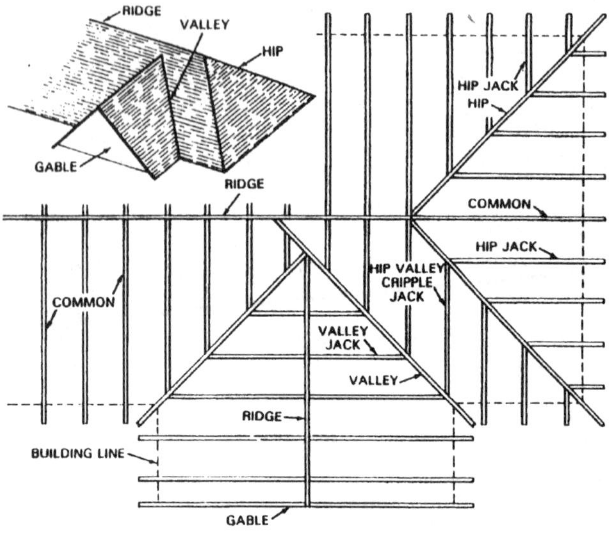

Figure 7-2 Roof Rafters

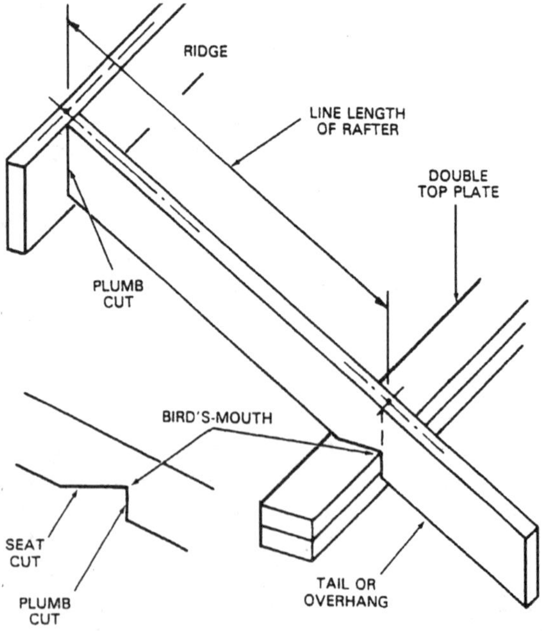

Figure 7-3 Rafter Detail

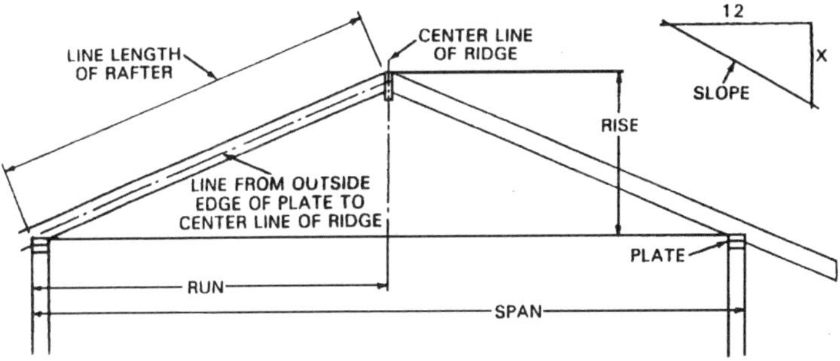

Figure 7-4 Rafters Cross Section

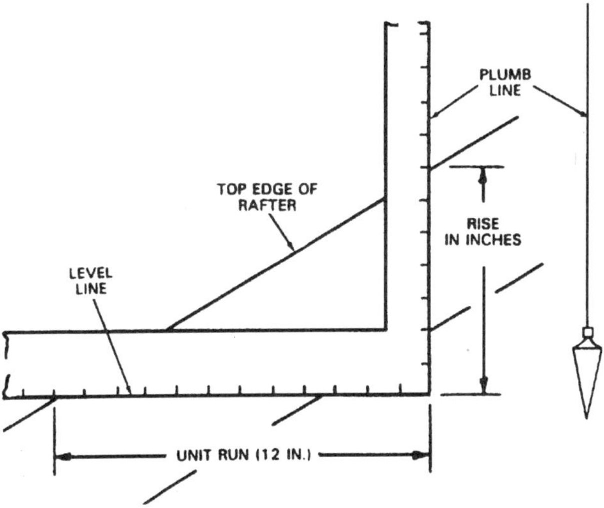

Figure 7-5 Rafter Layout

The actual length of the rafter is never figured, nor does it need to be—because all the measurements are made as horizontal and vertical distances with the aid of the square. Today, many framers prefer to do rafter layout with the aid of tables that give actual rafter lengths for various pitches and horizontal distances; these tables are stamped on the framing square itself or printed in pocket-size booklets that accompany the tool.

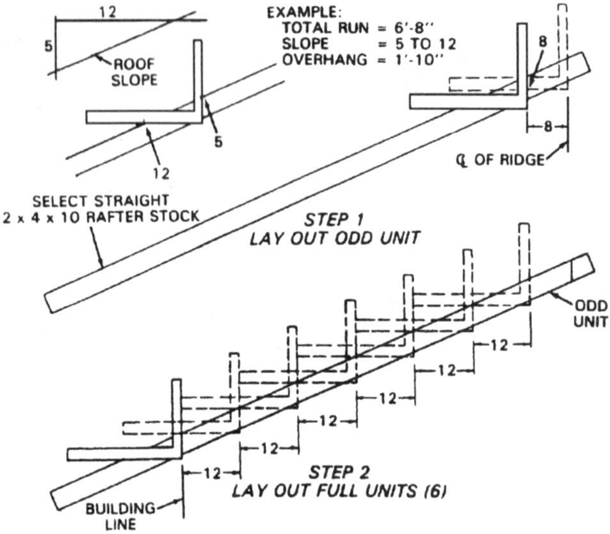

Figure 7-6 Rafter Layout

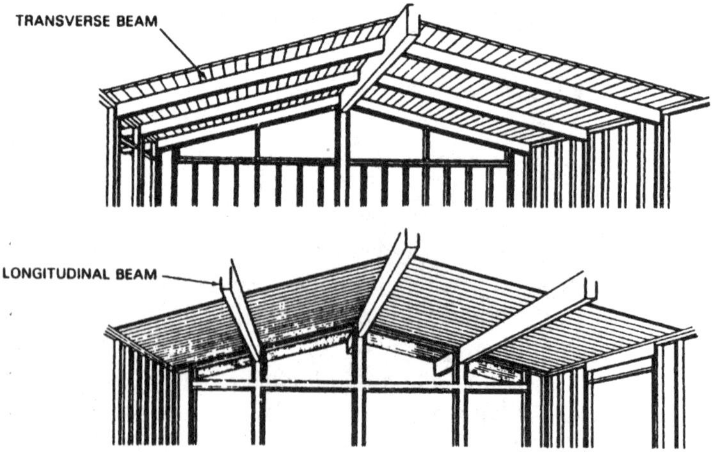

Figure 7-7 Post-and-Beam Rafters

Hips and valleys introduce another level of trigonometric complexity in rafter layout, but the experienced framing carpenter has little difficulty even here: Again, he or she can use published tables for hip and valley rafters, or do the layout the old-fashioned way.

The lead carpenter lays out only one rafter of each type by these procedures. He or she lays out the plumb cut and seat cut (known as the *bird's mouth*) and the pitch cut on one rafter. The carpenters go up the ladders to the ridge beam to see how it fits, minor adjustments are made to the stick, back up the ladders for a final check, then this stick becomes the *template rafter* from which other carpenters trace and cut the remainder of the rafters.

Roof Trusses

Standard roof trusses are lightweight prefabricated repetitive-member wood truss systems. Although wood trusses have been used for centuries, the more recent emergence of engineered, mass-produced, prefabricated, lightweight wood or wood-steel composite trusses has significantly altered the basic design philosophy of the construction industry. See Figs. 7-8 and 7-9.

Prior to the evolution of these lightweight truss systems, the wood truss was considered a relatively complex and

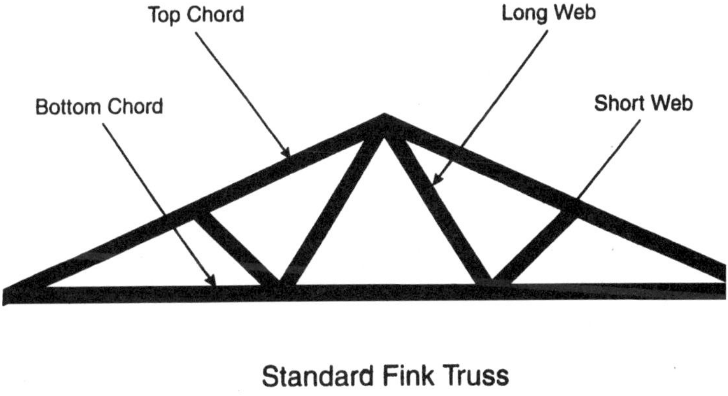

Figure 7-8 Roof Truss

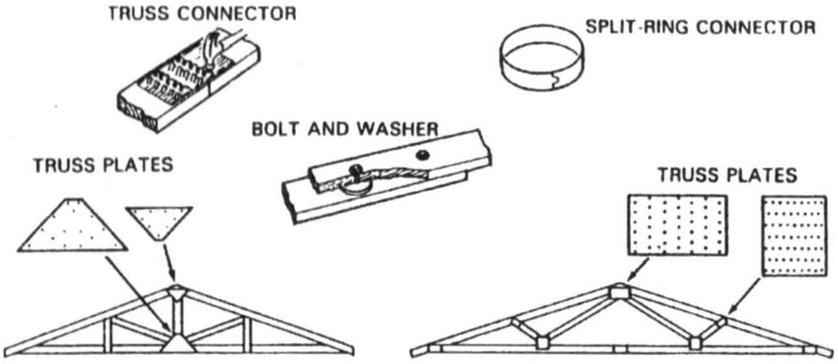

Figure 7-9 Roof Truss Components

labor-intensive structural component. Typically, wood trusses used in commercial construction were heavy solid sawn or glued-laminated timber trusses fabricated on site or in a laminating plant from detailed engineering designs.

Roof truss framing systems usually span from 12 to 20 ft (3.66 to 6.10 m) with dimension lumber purlins framing between them. Other relatively long clear-span roof framing systems which utilize timber trusses include trussed arches, portal frames, and rigid bents.

There are some common heavy timber truss framing configurations. These trusses can be constructed of individual planar components, i.e., all web members and chords occurring in the same plane, or can be of a multiple-member design, where the chord members consist of two pieces with the web members placed between them. In either case, the members can be joined by a variety of techniques. However, bolts or bolts used in conjunction with timber connectors (shear plates or split rings) with either steel or wood side plates are most commonly used.

The king- or queen-post truss forms a unique example of timber roof truss construction by combining a heavy timber compression chord with a steel rod or cable tension chord. These chords are customarily joined by vertical steel pipe webs.

Although most often associated with building construction, the heavy timber truss also has been used effectively in timber bridge construction. The early covered bridges commonly used some form of heavy timber truss as the main structural load-carrying element. The protection from direct exposure to the elements afforded by the external covering of these structures has allowed many of these trusses to perform for well over 100 years. Typical timber bridge truss configurations include parallel-chord trusses, bowstring trusses, pony trusses, and through trusses.

With the relatively wide spacing involved in these heavy timber truss systems, each truss was considered a primary structural element with little or no allowance for load sharing between trusses. Prior to about 1970, solid sawn heavy timber truss designs were based on working stress values extrapolated from the results of laboratory tests made on small clear samples of the wood species used in the truss.

As laboratory equipment capable of testing larger wood specimens was developed and research funding became available, a considerable amount of research was conducted to determine what effects load duration, moisture content, and the size of the member had on its ultimate strength (see Fig. 7-10). The results of this continuing research indicate that many of the working stresses used as a design basis for the large solid sawn timber trusses designed prior to these studies may have been nonconservative.

A consequence of this lack of qualitative information by early truss designers combined with little or no load sharing between trusses has been that some older heavy timber truss roof systems have shown signs of distress that must be corrected by field repair or reinforcement. It should be noted that many of these early trusses were designed for use in so-called temporary structures during the World War II era. Although these trusses were intended for a design life of

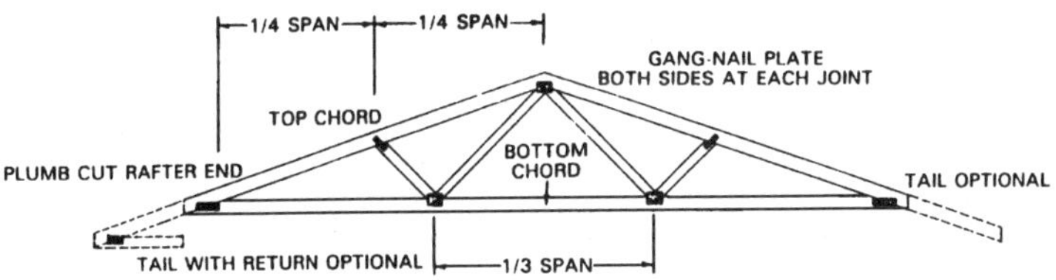

Figure 7-10 Truss Span Strength

less than 10 years, the fact that many of them are in service more than 50 years later is a testimony to the structural capabilities of heavy timber roof trusses.

In most cases it is difficult to identify the true cause of truss distress, since it may be related to moisture-content change and subsequent shrinkage stresses, excessive loads caused by a number of environmental factors, a change in building use, improper fabrication, long-term duration of load effects, high temperature as may occur in attic spaces, or other factors that would affect load-carrying capacity.

Time, experience, and considerable study have combined to provide a much broader knowledge of the failure mechanisms involved in roof trusses, thus allowing the framer to analyze these structural systems more effectively. With continued advancements in the state of the art of truss design, the evolution of improved methods of joint connection, and the development of new structural wood products, different forms of truss systems will remain a viable part of American wood framing standards.

In addition to the use of traditional forms of heavy timber trusses (see Figs. 7-11 and 7-12), it is inevitable that more sophisticated concepts such as space frame systems using heavy timber trusses will continue to evolve as architects combine the increased knowledge of the performance of individual truss components with CAD (computer-aided-design) methodology.

The introduction of sheet-metal gusset nail plates as a joint-connection device for lumber trusses combined with manufacturing technology capable of producing continuous truss chord elements using structural end joints has led to a new design philosophy in wood truss systems. Utilization of this technology allows the lightweight wood truss to be fabricated quickly and economically on a mass-production basis.

These lightweight wood trusses installed at relatively close on-center spacing and carrying correspondingly lighter loads could effectively match the allowable spans of their heavy timber truss counterparts. The framing system that eventually evolved from these developments consisted of

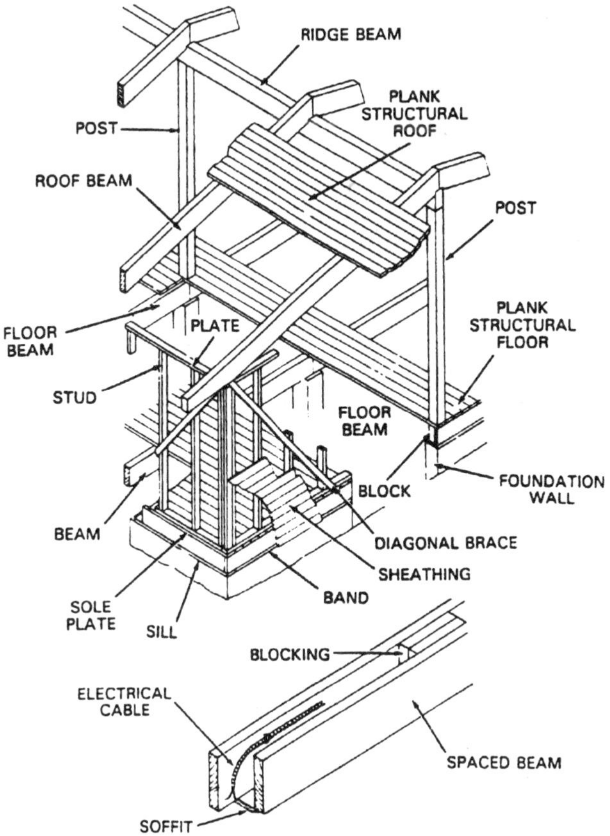

Figure 7-11 Heavy Timber Roof Construction

lightweight wood trusses placed 12 to 48 in (30.5 to 121.9 cm) on center with the decking material attached directly to the top chords of the trusses.

Lumber-trusses and lumber-metal composite trusses constitute the two categories of lightweight wood trusses that utilize this repetitive-member framing concept. Lumber trusses are constructed using dimensional lumber for the chords and webs. Basically, this connection involves the use of two sheet-metal gusset plates that are installed by physically pressing them into the opposite sides of a truss joint.

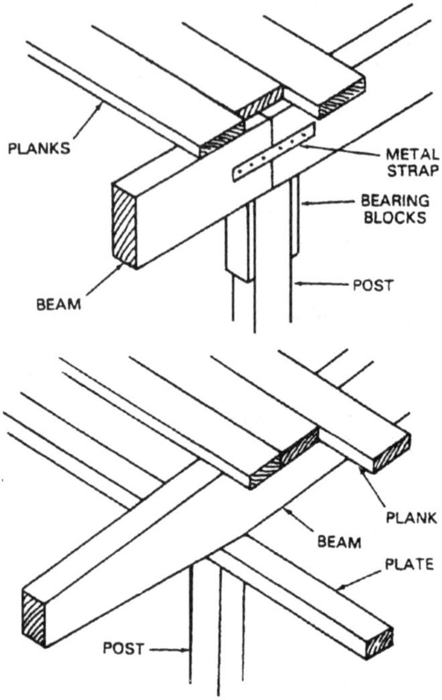

Figure 7-12 Post-and-Beam Roof Components

These plates are sized in such a way that they cover all the members to be connected at a joint with a sufficient area of metal gusset plate to effectively transfer the load from one member to another based on an allowable value per square inch of plate. Chord splices are accomplished by using two plates applied to the chords in the same manner, sometimes using reinforcing elements.

Metal gusset plates from one manufacturer are not generally interchangeable in a specific design with another producer's plate product. The plate's steel gauge and the number, size, and design of the steel projections (teeth) all affect its ultimate load transfer capacity. Most plate manufacturers provide engineering and computer truss design support for their product based on the code minimum roof classes shown in Tables 7-1 and 7-2.

Table 7-1

MINIMUM ROOF CLASSES

TYPES OF CONSTRUCTION							
I		II		III		IV	V
F.R.	F.R.	One-hour	N	One-hour	N	H.T.	One-hour
B	B	–	–	–	–	–	–
B	B	B	–	B	–	B	B
B	B	B	B	B	C	B	B
B	B	B	B	B	B	B	B
B	B	B	B	B	C	B	B
B	B	B	B	B	B	B	B
B	B	B	B	B	C	B	B
A	A	A	A	–	–	–	–
A	B	B	B	B	B	B	B
A	B	B	–	B	–	B	B
A	B	B	–	B	–	–	B
B	B	B	B	B	C	B	B
B	B	B	B	B	C	B	B
B	B	B	B	NR	NR	NR	NR
B	B	B	B	B	C	B	B
B	B	B	B	B	B	B	B
B	B	B	B	–	–	–	–
B	B	B	B	NR	NR	NR	NR

Table 7-2

WOOD SHINGLES	WOOD SHAKES
Not Permitted below 3 Units Vertical in 12 Units Horizontal (25% Slope)	**Not Permitted below 4 Units Vertical in 12 Units Horizontal (33.3% Slope)**
See Table 15-C	**See Table 15-C**
Shingles shall be laid with a side lap of not less than $1\frac{1}{2}$ inches (38 mm) between joints in adjacent courses, and not in direct alignment in alternate courses. Spacing between shingles shall be approximately $\frac{1}{4}$ inch (6 mm). Each shingle shall be fastened with two nails only, positioned approximately $\frac{3}{4}$ inch (19 mm) from each edge and approximately 1 inch (25 mm) above the exposure line. Stater course at the eaves shall be doubled.	Shakes shall be laid with a side lap of not less than $1\frac{1}{2}$ inches (38 mm) between joints in adjacent courses. Spacing between shakes shall not be less than $\frac{3}{8}$ inch (9 mm) or more than 5/8 inch (16 mm) except for preservative-treated wood shakes which shall have a spacing not less than $\frac{1}{4}$ inch (6 mm) or more than $\frac{3}{8}$ inch (9 mm). Shakes shall be fastened to the sheathing with two nails only, positioned approximately 1 inch (25 mm) from each edge and approximately 2 inches (51 mm) above the exposure line. The starter course at the eaves shall be doubled. The bottom or first layer may be either shakes or shingles. Fifteen-inch or 18-inch (381 mm or 457 mm) shakes may be used for the starter course at the eaves and final course at the ridge.
Per Section 1508.5 Per Section 1509	

Generally, metal gusset plate truss fabricators own franchises that allow them to produce trusses using the plate manufacturer's machinery, truss engineering service, and plates. While the plate manufacturer maintains quality control of the gusset plates, the overall quality of the truss is the responsibility of the fabricator depending on its internal quality-control requirements. The Truss Plate Institute (TPI) and the Wood Truss Council of America (WTCA) are the primary trade associations related to this industry.

In contrast to metal gusset plate trusses, composite wood-metal trusses are generally proprietary, with the manufacturer claiming exclusive rights to their production based on their uniqueness. One form of composite truss consists of wood chords and tubular steel webs joined together by simple steel pin connections. The wood chords of these trusses consist of either machine-stress-rated (MSR) dimension lumber structurally end-jointed to form continuous chords or laminated veneer lumber (LVL).

Another form of composite truss that is actually a variation of a metal gusset plate lumber truss is constructed of dimension lumber with either some or all of the webs composed of a steel angle with nail plates formed in its ends. These formed nail-plated angle webs are installed by pressing them into the opposing faces of the wood chords. The chords are spliced together in the same manner as are metal-gusset-plated lumber truss chords.

Because of the uniqueness of each form of proprietary composite truss, the manufacturers generally provide engineering support in the form of load tables and other essential technical information for the proper design of these truss systems. In addition, these national truss manufacturers commonly inspect and maintain internal manufacturing quality-control programs to ensure the structural integrity of these trusses.

While used primarily in applications with relatively close on-center spacings, i.e., 48 in (121.9 cm) or less, lightweight lumber or composite trusses also can be used in a girder truss framing system with some limitations. In order to use these trusses at on-center spacings of 8 ft (2.4 m) or more, it is common practice to tie two or more of them together in such a way that their individual capacities are additive, because the load factors differ from calculating common rafters as shown in Table 7-3.

You can combine multiple lightweight trusses to form a girder truss. The specific framing details will dictate the actual load distributions among the trusses thus joined. If these framing details apply loads to one side of the multiple trusses, it would be improper to consider equal load sharing. Experience has shown that three lightweight trusses are the maximum number that can easily and effectively be joined in this fashion.

Normally, multiple lightweight trusses built up as a girder truss are used in a floor or roof system where it is necessary to frame around openings created by skylights, chimneys, and stairwells or where a large concentrated load, such as from a rooftop unit, requires a greater truss capacity to carry the load. Multiple lightweight trusses tied together may not be as economical as a heavy timber truss because of the cost and difficulty associated with connecting them together, but it is a possible design alternative.

Truss System Loading

The minimum design loads to be resisted by any structural member or assembly are normally set forth by the governing

Table 7-3

Overhang	12"	16"	18"
Rise	3"	7"	4"
Run	6"	7"	6"
Span	12"	14"	12"
Pitch	1/4	1/2	1/3
Inches of Rise per Foot of Run	6"	12"	8"
Cut	12 / 6	12 / 12	12 / 8
LineLength per Foot of Run	13.42"	16.97"	14.42"
Total Line Length	79.77"	118.04"	85.77"
Total Rafter Length	93.19"	140.67"	107.40"
Purchase Length	94"	142"	108"

code authority, with the designer of record having the final responsibility of selecting the proper loads for each structural application. In addition to specifying these minimum design loads, the governing codes also indicate minimum combinations of loads to be considered for each design situation. While these loads and combinations of loads are common to all structural materials or systems, the manner in which they are applied or resisted may be unique to a specific material or system.

In wood truss systems, loads may be applied to the chords in a variety of ways, such as directly through the decking as uniformly distributed loads; as a series of relatively closely spaced concentrated loads, as from joists spanning perpendicular to the truss; or as concentrated loads from purlins spaced 48 in (121.9 cm) on center or greater.

The dead load imposed on a structural member is the weight of all the component building materials in a structural assembly supported by the structural member. The weight in pounds per square foot (newtons per square meter) of materials common to building construction can be found in various publications. While most dead loads are assumed to be uniformly distributed loads, loads imposed by mechanical systems are often applied as concentrated loads and should be applied at panel points wherever possible.

Interior partition walls or exterior walls bearing on the ends of cantilevers may result in additional uniformly distributed loads or concentrated loads acting on the truss depending on the positioning of the walls in relationship to the truss-framing system. Consideration must be given to proper detailing of roof-to-wall and floor-to-wall connections to ensure that nonloadbearing partitions do not become loadbearing walls as the trusses deflect under in-service loading.

The live loads imposed on a structural member are those loads that are due to the use and/or occupancy of the structure. Floor live loads are those due to human occupancy, furniture, materials stored on the floor, etc. The minimum recommended uniformly distributed design live loads for various uses and occupancies are listed in the governing building codes. In some applications, a wood floor truss may actually be subjected to several different uniform loads over a single span, such as a truss supporting an office loading over part of its span and corridor loading over the remainder of the span.

Similarly, trusses spanning continuously over two or more spans also must be evaluated for unbalanced loading acting on alternate spans in addition to full balanced loading on all spans to determine the maximum stress conditions for each truss component, i.e., webs, chords, and joints.

In addition to these uniformly distributed loads, the applicable code also may require that the floor be designed to carry a minimum concentrated load. This concentrated load must be considered to be positioned on a wood truss in a manner that will produce the maximum stress conditions in the truss members.

Roof live loads are those minimum loads established by the building codes to provide for temporary loadings that the roof must support, such as workers, equipment, and materials, in combination with roof framing computations as shown in Table 7-4. Roof live loads and snow loads are not considered additive; rather, that which produces greater stresses governs the design.

The subject of proper and adequate snow-load design for roof trusses is one that has received considerable attention. Snow loads on roofs can vary according to geographic location, climate, site exposure, shape and type of roof, whether the building is heated or not, and of course, from one year to the next.

Virtually all major building codes now recognize the need to design for these variable snow-loading conditions. Ground snow loads generally form the basis for the determination of the resultant roof snow loads, and the American Society of Civil Engineers publication ASCE 7-93, *Minimum Design Loads for Buildings and Other Structures*, provides ground snow loads by geographic area.

Of particular concern in the design of wood truss systems are drifting snow loads or accumulations such as may occur on a low roof located adjacent to the higher roof plane. This drift condition may result in a triangular load distribution being superimposed on the normal roof loading over a portion of the length of a roof truss spanning perpendicular to this high-low roof plane intersection.

For trusses spanning parallel to the high-low roof plane intersection, the resultant additive load is a uniformly distributed load with a maximum intensity for the truss located closest to this roof plane interface. In either case, the wood trusses must be designed to support the governing maximum loading condition with all truss components sized accordingly.

Table 7-4

Specs.	1	2	3	4	5	6	7	8	9	10
Rise	3'	7'	5'	8'	6'	4'	6'	3'	10'	4'
Run	12'	7'	20'	16'	18'	6'	8'	6'	24'	12'
Span	24'	14'	40'	32'	36'	12'	16'	12'	48'	24'
Pitch	1/8	1/2	1/8	1/4	1/6	1/3	3/8	1/4	5/24	1/6
Inches of Rise Per Foot of Run	3"	12"	3"	6"	4"	8"	9"	6"	5"	4"
Cut	3/12	12/12	3/12	6/12	4/12	8/12	9/12	6/12	5/12	4/12
Line Length Per Foot of Run	12.37 in.	16.97 in.	12.37 in.	13.42 in.	12.65 in.	14.42 in.	15.00 in.	13.42 in.	13.00 in.	12.65 in.
Total Line Length In Inches	148.44 in.	118.79 in.	247.40 in.	214.72 in.	227.70 in.	86.52 in.	120.00 in.	80.52 in.	312.00 in.	151.80 in.
Total Rafter Length w/ $1^1/_2$" Ridge and 1' Overhang	160.06 in.	135.01 in.	259.02 in.	227.39 in.	239.60 in.	100.19 in.	134.25 in.	93.19 in.	324.25 in.	163.70 in.

Of a similar nature is snow accumulation or ice buildup in valleys of adjacent pitched roofs, which can result in other than uniformly distributed loads acting on the roof trusses. Unbalanced snow loads required by the codes in the case of wood truss applications involving multiple-span members must be considered because they may create higher stresses or possible stress reversals in the truss elements and their joints.

Unbalanced snow loads can occur in other ways, such as on a cantilevered truss used to create a roof overhang or in a gable roof application where snow may accumulate on only one side of the roof peak. The effects of unheated attic spaces or highly insulated roofs also should be considered because they can result in uneven snow accumulations. ASCE 7-93 provides detailed explanations of snow load analysis for virtually all roof conditions and configurations that may be encountered.

The application of wind loads on the exterior surfaces of a building causes wind pressures and suctions on the building, which result in tension forces being imposed on the structural members. Considerable research has been conducted to evaluate wind effects on various structures, and this has resulted in the establishment of design pressure coefficients that account for building shape and wind direction.

Each of the major building code agencies has addressed the issue of wind design by providing various minimum wind pressures for a given geographic location on the basis of historical recurrence and the height above the ground. Wind-load pressures and suctions can have the effect of imposing uplift loads on roof trusses. In lightweight wood truss systems, the net uplift loads (uplift load minus the structure dead load) may result in stress reversals of the truss components and uplift reactions at the bearing locations. Similarly, wide overhangs created by cantilevered roof trusses may result in negative reactions at bearings, with wind loading causing possible stress reversal in web and chord members.

Seismic stresses are treated as the result of laterally applied horizontal forces (and vertical forces to a lesser extent) applied at each floor or roof level above the base.

These forces are assumed to come from any horizontal direction so as to create the maximum loading effect on the structure. As with wind loads, a particular concern in wood truss systems is that the connection between the truss and its supporting structure is capable of resisting the lateral forces induced by seismic deformation.

Roof Truss Erection Loads

A common problem related to the installation of roof trusses is the lack of proper consideration given to erection loads and the effects of these loads on the trusses. A typical construction practice is to place partial or complete bundles of sheathing material on trusses after they have been erected but prior to the complete installation of sheathing.

These erection loads can result in relatively significant top chord compressive stresses that can induce buckling of an improperly braced compression chord.

These erection stresses must be accounted for by using an adequate lateral bracing system, ensuring that it is properly installed during the erection sequence, and clearly defining the maximum allowable construction loads that can be placed on the truss system. Care also must be given to the methods used to lift the trusses into position so as not to overstress the unbraced truss compression chords or to create excessive stress reversals in the truss members.

Ponding

In the framing of flat or nearly flat roofs, special consideration should be given to the potential for ponding of rain water. Roof collapses that were directly attributed to water ponding on the roof have occurred, involving virtually all types of building materials.

When ponding of water occurs, for whatever reason, deflection is initiated and may continue at a progressive rate because the work energy of the ponding water is initially greater than the resistance of the truss. This deflection continues until equilibrium is reached between the added water load and the truss resistance or until failure occurs.

Ponding is usually a greater problem for roofs with long-span structural members than for roofs with short-span trusses, since the relative stiffness of a long-span system is likely to be less than that of a short-span system. Roofs with large areas tributary to a primary flexural member (truss or beam) are, in general, more susceptible to ponding than roofs with small areas tributary to a primary flexural member.

Ponding problems are not necessarily limited by geographic location, although the frequency of occurrences seems to be higher in areas where relatively low design live loads are used. Failures have occurred in semiarid regions, since rainstorms of high intensity may occur, even though the annual rainfall in these regions is small. Although roofs in colder climates are generally designed for relatively high live loads which increase the stiffness of the system, unusual weather conditions can occur that may result in the blockage of drainage paths due to packed snow or ice. In such cases, the ponding load becomes additive with the existing snow or ice on the roof.

While actually a part of the overall building design rather than a loading provision, providing an adequate slope to roofs—whether by cambering the truss, sloping the top chord, or varying bearing elevations—and making provisions for an adequate drainage system helps minimize most potential ponding problems.

While flat roofs can be designed to resist overloads due to ponding, this requires very stiff members and hence larger members than would be required if ponding were not allowed to occur. Such roofs may be less economical, and the framer must exercise the utmost care in structural integrity and control in the overall truss installation process to adequately allow for any intentional ponding.

Truss Systems Checklist

Transferring an architectural concept into reality challenges any building designer to determine the most efficient and cost-effective means to cover the required floor and roof spaces while maintaining the desired architectural features. A number of design considerations must be evaluated to accomplish this task using a wood truss framing system. While these considerations and cost factors are often interrelated, they individually affect the overall cost-effectiveness of the roof truss system.

The decisions facing the designer multiply in repetitive-member lightweight prefabricated truss applications. Each decision-making step must be evaluated not only individually but also with respect to its effect on other design variables as it affects overall system costs. The primary design parameters to be evaluated with respect to their structural efficiency and cost-effectiveness in a lightweight wood truss system are as follows in this checklist:

- ☐ *The design loads and the deflection requirements of the truss system:* While the minimum design loads and deflection criteria to be applied in a specific design situation are normally specified by the governing building code, the designer may want to consider more stringent load and deflection criteria.

  While designing to these more restrictive criteria may result in an increased system cost, the possible benefits of improved floor performance may justify this added cost.

- ☐ *The maximum desired clear span of the wood truss system:* It is generally recognized that shorter-span systems result in less overall cost than those with relatively long spans. As design spans decrease, lighter and shallower trusses or deeper trusses spaced farther apart may be used with an attendant cost savings to the system.

- ☐ *The maximum spacing of the wood truss system:* Generally, an increase in truss spacing results in cost savings because of the requirement of a minimum number of trusses. However, the maximum spacing may be dictated by other considerations. As truss spacing increases, the thickness of decking required to span between trusses increases, with a corresponding increase in deck costs.

The choice of ceiling construction also may control the truss spacing. In the case of a conventional suspended ceiling, a truss spacing of 48 in (121.9 cm) on center or more can be accommodated. However, if the framer intends to use a direct applied drywall ceiling, a code maximum spacing of 24 in (61.0 cm) on center is dictated. The cost and architectural consideration of the ceiling construction thus become an integral part of the overall wood truss system costs.

- *The depth of the wood truss system:* For a given span, spacing, and loading condition, generally as the depth of the truss increases, so does its ability to carry heavier loads. However, this increased depth may require higher support walls to be constructed to maintain interior height clearance requirements, thus adding additional building costs.
- *The locations of interior loadbearing elements:* The positioning of interior loadbearing walls or columns also can affect the wood truss system. Interior loadbearing columns supported by the wood truss system can result in relatively high concentrated loads being applied to the trusses and may require multiple trusses designed as a girder to support this load.

  Similarly, loadbearing walls supported by the wood truss system may require additional trusses if the walls are positioned parallel to the truss span, or stronger trusses positioned at a closer on-center spacing may be required to support these walls if they are placed perpendicular to the truss span. In either case, the supporting truss system has higher costs associated with it when required to support these loads, and alternative framing schemes not requiring such support by the trusses should be evaluated to optimize the structural system cost.
- *The routing of mechanical systems through the truss support structure:* The extent of mechanical systems that are located within the truss structure may require deeper trusses and larger panels or openings than would be needed to carry the actual design loads. In some instances,

the mechanical system layout may have an influence on the actual direction in which the trusses are to span, thus affecting system costs.

- *Foundation and wall support consideration:* While conventional wood or steel stud walls or masonry walls and their normal wall foundations are usually adequate to carry the loads imposed by uniformly spaced lightweight wood trusses, the use of widely spaced trusses, such as in girders trusses, may result in concentrated end reactions that require separate columns in the wall construction and associated heavier footings.
- *Rooftop or suspended mechanical units:* The placement of mechanical units also can influence the design and cost of the truss systems. These units, either placed on the roof or suspended from the trusses and located near the midpoint of the truss span, may require additional trusses for support or a stronger and thus more costly truss design. Placement of these loads near bearing walls may permit the use of the truss systems as designed for normal uniformly distributed loads with no increase in truss size or quantity.

It becomes readily apparent that the interrelationship of these design considerations makes any general conclusion regarding the economic advantages of one truss system over another virtually impossible.

However, the manufacturers of the lightweight wood and wood-steel composite trusses have developed a wealth of technical information in the form of design load tables, which greatly assist the design architect. In these tables the depth of the truss, the span, and the allowable load the truss can carry while meeting specified deflection criteria are given.

Combining these types of data with the maximum allowable spacing of the trusses arrived at from other design considerations and applying the relative costs of various decking materials allow the designer to establish the total cost of a wood truss system that achieves specific structural and architectural criteria.

It is suggested that the manufacturers of the engineered, lightweight prefabricated wood trusses be contacted in the conceptual stage of a project for assistance in making economic assessments of various possible framing schemes, as many of them have established so-called best systems for given design criteria.

Roof Truss System Performance

Satisfactory performance of a wood roof truss system is dependent on its ability to (1) resist all design loads that may be applied during the life of the structure with sufficient consideration of possible overloads; (2) provide adequate stiffness to maintain both short- and long-term deflections within acceptable limits; and (3) minimize movements, vibrations, and noise that may be objectionable to human occupants.

While minimum design loads are set forth by the governing building codes, the final design loads for each specific structural application are established by the designer based on experience and judgment. These specified design loads thus become the maximum anticipated loads to be carried by the wood truss system during the life of the structure.

Resistance to overloads or unanticipated loads thus becomes a factor-of-safety consideration. The tabulated design values set forth in the referenced design standards for the various materials used in a wood or wood-steel composite truss incorporate specific factors of safety. However, because of the interaction of the various materials and connection details used in an individual truss combined with effects of decking, bridging, and bracing, it is difficult, if not impossible, to assess the factor of safety of the total system and thus its inherent capability to resist overloads.

In the case of prefabricated lightweight wood trusses, the manufacturers gain acceptance through the governing code authorities based on a thorough review of test data and design evaluation. Inherent in these data are appropriate factor-of-safety considerations for the truss component materials and the truss itself. Integral to these code acceptances are the requirements for internal quality-control programs to ensure structural integrity. In the case of metal gusset plate lumber trusses, the code acceptances are limited to an evaluation of the plate connections only and not the fabricated truss, which compounds the definition of a factor of safety for the truss itself.

Generally, a system of relatively closely spaced trusses, i.e., 24 in (61.0 cm) or less on center as used in a repetitive-member system, is accepted as performing more effectively than an individual truss in resisting loads due to the inherent load-sharing aspects of the system itself.

In addition, the possibility of a weak member occurring in a given truss system is offset to a degree by the probability that a stronger member will be located adjacent to it, thus helping to resist the imposed loads. In the case of an individual heavy timber truss application, additional concern in the selection of maximum design loads may be warranted, since system overload redundancies are less likely to be present than in a closely spaced truss system.

Short- and Long-Term Deflections

A wood truss system deflects under loading, with the deflections being time-dependent relative to the creep and set characteristics of the truss components. The proper design of wood truss systems should limit these short- and long-term deflections to acceptable bounds through material selection, connection design, and truss proportions. In addition, camber provisions should be incorporated in the truss design to account for these anticipated deflections.

Deflection is defined as the vertical movement of the truss under load, whereas *camber* is that vertical curvature manufactured into the truss to offset the calculated deflections and thus eliminate possible sagging of the truss below a level line under service loads.

Creep is the movement or deflection that occurs over time under continued loading. *Set* is the permanent deflection that remains after loads are removed and is caused by inelastic deformations between the wood fibers and slip or yielding at the truss joints.

The deflection limits for wood truss systems set forth by various building code authorities and standards are separately specified for roof and floor systems. The limits for roof systems generally are directed to ensure adequate drainage with additional consideration for possible distress of ceiling materials that may be directly applied to the truss chords.

Typically, roof live-load deflections are limited for those applications receiving a plaster ceiling and for designs requiring the use of more flexible ceiling coverings such as drywall. In most cases, this deflection limit is set when the overall building design provides a positive roof slope for drainage and where this amount of deflection will not cause distress to the roof construction itself.

In addition to these general deflection criteria, designers also should evaluate roof truss deflections for special conditions such as a truss system supporting a glass window wall or skylight, which may require highly restrictive deflection limitations. Roof trusses supporting movable or folding partitions generally require more restrictive deflection limitations to ensure the long-term operation of these partitions. Another example where deflection may be of primary design importance would be in the design of a flat or nearly flat roof where ponding of rainwater may occur and a relatively stiff system is warranted. Overall evaluation of the deflection characteristics of the truss system also should take into account the deflection of the decking material spanning between trusses.

In heavy timber truss applications, the primary deflection concerns are to minimize distress to other roof construction materials and to eliminate detrimental visual effects that would be associated with an appearance of sagging or deflection of the bottom chord below a level plane. It is essential to appraise the degree of slip or set in the joints of any wood truss using mechanical connection devices, in assessing the overall deflection performance of the truss.

Camber of wood roof trusses is often provided to assist in ensuring proper drainage, for appearance, and to offset creep of set due to long-term loading. Specification of camber must be carefully evaluated to ensure that it does not cause framing problems as may occur at a mismatch of adjoining roof planes or the inability to create a straight fascia line at a building overhang. A commonly used guideline for lightweight, repetitive-member truss systems is to provide camber equal to $1^1/_2$ times the calculated dead-load deflection to allow for both initial set plus long-term creep deformation. Another suggested camber policy often used is to camber for dead-load deflection plus one-half the live-load deflection. The final responsibility for proper roof truss camber lies with the building designer.

Floor truss systems require more restrictive deflection limitations based on the desirability of stiff floors as a human response issue. A live-load deflection limit of $^{30}/_{360}$ has been universally applied for many years. However, for the longer spans achieved using lighter truss systems that have evolved through advancements in wood truss technology, this criterion becomes questionable.

A suggested deflection criterion based on occupancy satisfaction for these longer spans, beyond 20 ft (6.1 m), is to limit live-load deflections to $^{40}/_{480}$ for residential floors and $^{58}/_{600}$ for nonresidential floors. While this type of criterion has not been adopted by the governing building code authorities, it has been applied to numerous building applications with apparent success in minimizing objectionable floor deflections.

A composite action between floor sheathing and the truss system is achieved through proper nailing schedules and to a greater extent by the addition of a field application of elastomeric adhesive as a connection between the sheathing and the trusses. Some prefabricated wood truss manufacturers have conducted tests of this composite action and established composite-action design procedures accounting for this additional deflection resistance. Elastomeric adhesives have generally been limited to floor-system applications, since they are not considered effective in reducing roof deflections because of their thermal creep characteristics.

As with roof truss systems, the possibility of creep or set should be considered in floor-system design, with particular attention to the potential deformations associated with mechanical joint connections such as may occur in a metal gusset plate truss. A floor truss system that takes a significant set over time can be objectionable in appearance and may cause separations at partition moldings, cracks in interior walls, or possible distress in ceiling construction.

For repetitive-member systems, both roof and floor, it is advisable to provide additional trusses to support concentrated loads such as may be caused by the installation of heating and ventilating equipment and by partitions to equalize deflections of the truss system.

The camber of wood floor truss systems is somewhat less critical than that of roof trusses, since drainage is obviously not a concern and, in general, these trusses are inherently stiffer than roof trusses because of their more restrictive deflection criteria and the generally higher loads used in their design.

A recommended camber policy is to provide floor trusses with a camber equal to $1^1/_2$ times the anticipated dead-load deflection, thus accounting for initial deflection and long-term creep from permanent loads. If the floor system is more likely to sustain heavy long-term live loads such as in warehouses or libraries, it is recommended that camber requirements be increased accordingly.

Truss Bearing and Anchorage

Wood floor and roof truss systems must be designed to perform in an integrated manner with other structural elements of the overall building system. Provision for support of the truss system, adequate anchorage, lateral support, and bracing of these structural elements must be evaluated within the overall scheme of stability and structural integrity of the total framing structure.

Because buildings are so varied in size and shape with a virtually infinite array of structural framing schemes possible, these specific support requirements and bearing details for wood truss systems must become the responsibility of the building designer.

Essential to the performance of wood trusses is adequate support and attachment at the bearing locations. Inadequate attention to bearing details is probably the greatest factor associated with the failure of lightweight prefabricated repetitive-member wood truss systems. These support and anchorage inadequacies may be categorized into three problem areas:

- Insufficient capacity of the support structure to resist truss reactions
- Insufficient bearing between truss and support members
- Improper attachment details, particularly with respect to uplift forces

Perhaps the technological developments achieved in the design and fabrication of repetitive-member lightweight prefabricated wood truss systems have not had matching developments relative to the structural support of the products. These lightweight trusses are often utilized in relatively long span applications, creating greater reactions than would be realized with more conventional 2×8 or other framing systems.

One result has been structural inadequacies in ledger support details and attachments and insufficient support capabilities of walls, headers, and beams. This problem becomes most evident in situations where conventional framing methods have been extrapolated routinely to accommodate the longer spans without adequate engineering verification.

Prefabricated trusses are normally manufactured to a predetermined specified span dimension. So it is imperative that the supporting structure be constructed to tolerances that match the span dimensions of this prefabricated truss. If these dimensional requirements are not met in the field, concessions may be made or poor judgment used regarding bearing modifications, resulting in inadequate bearing support for the trusses. Field inspection by a qualified manufacturer's representative, building inspector, or the design professional is thus necessary to ensure bearing support adequacy.

Lightweight wood roof truss systems are susceptible to the effects of net uplift forces that may develop during periods of high winds. Inadequate bearing attachment to resist these forces may result in either folding one bay of trusses over the next bay or simply lifting an entire bay off the supporting structure. The applicable codes establish minimum design wind uplift forces, and proper structure design should include adequate anchorage of the trusses at their support to resist these reactions. This anchorage requirement must be carried to total structure resolution.

Some manufacturers of prefabricated lightweight wood trusses have developed standard bearing details with accompanying load capacities as an aid to the designer in ensuring that adequate bearing provisions are provided for all possible design loads.

Lateral Support Considerations

The required lateral support to resist buckling of the compression chords of lightweight repetitive-member wood truss systems is most often provided by the sheathing material, which is directly attached to the top chord members of the truss. While the top chord is generally the compression chord of a truss, special consideration needs to be given to provide adequate lateral bracing to the bottom chords of trusses for cantilever conditions, for trusses spanning continuously over two or more spans, or when wind loading causes a net uplift loading to the truss, since under these varied conditions, the bottom chord becomes the compression chord over part or all of its length.

A plywood or other wood-based panel product acting as a structural diaphragm provides excellent lateral support for the trusses using simple mechanical connection to the top chord members. The addition of an elastomeric adhesive in addition to the mechanical fastening of the panel product to the top chords further enhances this lateral support. Other types of decking or cladding materials also can be designed to provide this lateral support. Solid sawn or laminated tongue-and-groove decking of various sizes can provide effective lateral support to the top chords.

However, it is essential that in addition to the attachment of the decking to the trusses, care be taken to ensure that each course of the decking be attached to the adjacent course such that an integral diaphragm is formed. Again, application of an adhesive to the tongue-and-groove joint of this decking can significantly increase its diaphragm resistance values. Reference may be made in the adopted building code for specific truss diaphragm values for a nailed-glued-laminated timber deck system.

Joists or stringers framing across or between the top chord members of the trusses at relatively close on-center spacing with a decking material attached directly to these subframing members also can provide sufficient lateral bracing. This requires that the deck itself provide suitable diaphragm stability and that the attachment of the stringers to the trusses be adequate to provide the top chord restraint to the extent presumed in the truss design.

The use of heavy timber decking or purlins spanning between trusses is quite common in roof framing designs using heavy timber trusses, which generally have wide on-center spacings. However, heavy timber decking and/or purlin framing also can be used in systems utilizing lightweight prefabricated trusses spaced at 6 to 8 ft (1.83 to 2.44 in) on center or with multiple trusses tied together to form girder trusses at greater on-center spacings. Regardless of the type of truss, the requirements for adequate bracing of the compression chords are the same.

As general guidelines, when heavy timber decking is nailed directly to the truss chords or when roof joists are set between the top chords with the decking nailed directly to the joists and chords, the least dimension of the top chord is taken as the depth of the chord. Similarly, when roof joists are set on top of the compression chord and securely attached to the chord and to blocking set between the joists with an adequate diaphragm attachment to the joists and blocking, the least dimension can again be taken as the depth of the chord.

For other conditions, the least dimension should be taken as the width of the top chord. When decking or blocking between roof purlins does not provide continuous lateral support to the top chord, the unbraced length for determining lateral buckling characteristics should be taken as the spacing between purlins. The truss should then be analyzed as a beam column in accordance with the provisions of the *National Design Specification for Wood Construction*.

While sometimes used in conjunction with wood roof trusses, corrugated metal decking may not provide adequate lateral restraint to the top chords. The nailing or screw attachment to the truss is normally made through the top of the corrugation rather than at the valley to enhance watertightness. The inherently longer nails or screws required by this connection detail do not necessarily provide the required lateral support to the truss. In addition, corrugated decking in itself may not provide required diaphragm resistance. A preferred detail is thus to provide a light plywood or wood panel sheathing material under the corrugated decking.

Various "composite wood fiber" decking materials are available that may be used in conjunction with wood truss framing. While these materials provide excellent vertical-load-carrying capacity, their ability to transfer lateral loads is significantly less than that of other common wood deck products. Some of the manufacturers of these products have performed diaphragm testing and provide recommended attachment details with corresponding load capacities.

However, numerous lateral stability problems have been associated with wood truss systems using this type of deck because of improper or insufficient mechanical attachment of the deck to the truss chords. The requirement of relatively large nails or spikes to penetrate these thick—2 in (5.08 cm) or more—decking materials also causes potential splitting problems in the wood chords. The designer is thus cautioned that care be given to attachment detailing and required diaphragm strengths when using the wood fiber deck products in conjunction with wood trusses.

Lateral support requirements provided for lower chord members when loading in compression are additional design considerations. Directly attached ceiling or soft materials may not be sufficient to provide adequate lateral restraint, and other methods must be considered by the designer.

Bridging and Bracing

Bridging in truss systems is installed to provide vertical-load sharing between trusses. Bracing in truss systems is used to provide lateral stability and alignment to the system or its component members, either as temporary bracing during construction and/or as a permanent bracing requirement. With proper provision, temporary bracing may fulfill some of the bracing requirements.

The purpose of bridging differs for roof and floor systems. In roof systems, real loadings tend to be somewhat uniform or uniformly varying, with obvious exceptions such as rooftop heating and ventilating equipment or the occasional maintenance personnel loadings. Thus bridging in roof systems is primarily provided to give structural redundancy to the system; i.e., a weaker truss is partially supported by stronger trusses adjacent to it.

The extent of load transfer to be provided with wood roof truss systems is not clearly understood. In some cases, an ultimate load transfer required per bridging line is specified with a maximum spacing of bridging lines given. Another philosophy of viewing load transfer requirements is to require greater load transfer capability for truss systems that have less reliability.

For example, a truss system of relatively low grades of visually graded lumber manufactured with minimal quality-control provision should have a more extensive bridging system than a truss system utilizing a high grade of machine stress-rated lumber fabricated under a more restrictive quality-control system. This essentially indicates that the more reliable and predictable the truss, the less the redundancy is needed.

It is relatively easy to provide effective bridging and thus load transfer to trusses that are spaced closely together. It is more difficult to provide effective bridging to trusses which are spaced 8 ft (2.44 m) or more on center. Hence the building designer generally considers these trusses to have little redundancy with respect to load sharing.

The load-sharing requirements between trusses of a floor system have greater significance in transferring concentrated loads than with a roof system. The purpose or objective of floor system bridging tends to shift to floor performance regarding deflection and vibration characteristics rather than being a purely structural concern. Again, there are few guidelines for establishing load transfer requirements. The typically closer on-center spacing of floor trusses makes it easier to achieve effective load transfer, and if it can be demonstrated that sheathing applied directly to the top chords provides significant transfer, acting as a tension tie between trusses.

Bridging details can be of several types. Strongback bridging is accomplished by extending a light solid sawn joist, typically 2 × 4 or 2 × 6 in, through the trusses at a panel point location of either the top or bottom chord with some form of mechanical attachment between the bridging and the chord. Metal or wood crossbridging is often used, but the attachment detail becomes quite critical because slip at the mechanical connectors can render this type of bridging virtually ineffective.

A combination of strongback and crossbridging has been utilized for deeper trusses with greater on-center spacings. The strongbacks are typically provided at both top and bottom chord members, with crossbridging then installed between these strongbacks. Another type of bridging used that is generally less effective is to provide a flat 2 × 8 member laid across the top of the bottom chords and mechanically attached thereto to act in tension as the truss deflects.

The permanent bracing required for wood truss systems can satisfy two purposes: (1) bracing of the total structure, which incidentally is located within the truss system (such as with seismic ties or diaphragm bracing) and (2) bracing to stabilize the individual trusses or their components. As discussed previously, this latter bracing type is provided primarily to stabilize the compression members of the trusses, including web members acting in compression, or to stabilize the bottom chord tension members from lateral movements.

This bottom chord stabilization is recommended when a directly applied ceiling is not utilized and is highly beneficial where eccentricities exist in the panel point connection of the first bottom chord panel point of top-chord-bearing trusses.

The sizing and connection requirements for bracing systems are highly judgmental. Standard wood engineering design principles are applied in this design as guidelines for ratios, etc. A load criterion often referenced is to provide lateral support members capable of resisting 2 percent of the compressive force acting in the compression chord.

This rule applies when the chord members are in true alignment. If the members are misaligned or otherwise deviate from a straight line, the additional eccentric forces generated must be resolved. The frequency with which this lateral stabilizing is to be provided along the compression chord is left to the judgment of the designer. However, it is obvious that the chord member must have adequate buckling stability between supports to carry the design loads.

Analysis and Design of Wood Trusses

Because of the wide variety of wood materials and related technology available, there are virtually unlimited variations of profile and capacity in today's wood trusses. Because of the wide range of possibilities, the often proprietary nature of light-truss standards, and ready availability of timber truss standards, a complete coverage of wood truss design procedures is not practical in this handbook.

It is of course possible to design wood trusses based on sound engineering practice, with reference to wood design and connector values such as those found in the *National Design Specification for Wood Construction* and elsewhere and to the applicable standards of steel and other component materials. The wood truss manufacturer considers connection details most carefully, since, as in most types of structures, it is in such details that design errors leading to in-service problems most frequently occur.

Design Standards

Lightweight factory-manufactured trusses are designed and produced in accordance with standards that are often wholly or partially proprietary in nature. The metal gusset plate truss industry has two general design standards promulgated by the building codes. These are the *Design Specification for Metal Plate Connected Wood Trusses* and the *Design Specification for Metal Plate Connected Parallel Chord Wood Trusses*. The former is applicable only to pitched truss designs, while other more complex configurations must be designed under standards established by a particular plate manufacturer.

Large sawn timber or glulam trusses are often designed in accordance with standards found in the *Timber Construction Manual* of the American Institute of Timber Construction. Those desiring further information on proprietary design methods should refer to manufacturers' literature or make direct inquiries to the engineering staff of such organizations.

Typical Spans and Proportions

The common prefabricated lightweight wood trusses typically span from 15 to 50 ft (4.57 to 15.24 m) at spacings from 12 to 48 in (30.48 to 121.92 cm) on center. However, trusses of this type capable of much greater spans and of wider on-center spacings are available from some manufacturers. There is no span-depth ratio limit for these trusses; rather, their dimensions are controlled by the strength and stiffness of the materials and connections used.

The range of span-depth ratios typically encountered for prefabricated lightweight wood trusses is given in code tables. However, an example of design innovation is a relatively light trussed arch of parallel laminated veneer lumber chords combined with tubular steel webs used to clear-span 400 ft over the football stadium of the University of Idaho.

Heavy timber trusses, including those fabricated from either solid sawn timber or glued-laminated members, have been used in spans of 150 ft and more in both building and bridge applications. In residential and light commercial framing, such trusses are normally spaced from 10 to 25 ft (3.038 to 7.62 m) on center. The profiles of heavy timber trusses are typically determined by consideration of appearance, clearance, and overall structure configuration. Experience has shown that for common profiles, the greatest design efficiency and economics are attained when span-depth ratios are limited approximately to the ranges listed in the code tables.

Preliminary Design

Considered here are common types of trusses used in buildings only, since applications such as bridge trusses, long-span trussed arches, or other complex applications require additional considerations unique to the particular project and are beyond the scope of this handbook. The design architect must first have an awareness of products, materials, and manufacturing facilities available to service the project. In the case of long spans or deep profiles, investigation into possible shipping problems is advisable.

When shipping of the trusses as a preassembled unit is not practical, members must be assembled on the job site, which requires considerations of available space, proper scheduling, and perhaps the practicality of field assembly of certain details. For example, a glued splice may require controlled temperature and pressure, which are difficult to achieve in field conditions.

Whatever the type of truss considered, the best sources of design information for estimating are those who supply, design, and/or fabricate wood trusses. If information is not at hand, a few telephone calls generally secure the desired literature. Many suppliers, manufacturers, and fabricators provide preliminary design and cost-estimating services.

Most manufacturers of prefabricated lightweight trusses publish literature that includes load-span tables, truss dimensions, and typical details that can be used for layout. Special profiles and associated details are often available on request.

When it is desirable to determine timber truss sizing on a preliminary basis, the following procedure can be used, but the results give only an approximation of required profile and member sizes.

Typical Analysis of Wood Trusses

Discussions of the truss as a structural member abound in texts of structural analysis. Of course, the pure theoretical truss with loads applied only at frictionless pins is as difficult to achieve with wood as with other materials. However, when certain design principles are followed, such as symmetry of joint fasteners about the centerlines of the member, tests and experience have shown that results are satisfactory.

Therefore, wood truss analysis usually proceeds with a theoretically pure truss analysis coupled with an evaluation of secondary stress in individual members and design of the connections. It is the orthotropic nature of wood and special connections that have evolved that leads to somewhat unique considerations in wood truss design.

As noted, the first step in wood truss analysis is usually the assumption of a simple planar truss with loads applied at panel points. Pure axial member forces can be determined by the method of joints, method of sections, graphic analysis, or use of one of the many computer programs available, with truss-type members specified. Bending moments are added to axial forces when the member is loaded transversely.

Probably the most commonly used procedure for determining truss member forces by manual calculation is the *method of joints*. This method of analysis, which is described in detail in numerous textbooks on structural analysis, lends itself well to trusses that are not very complicated, such as simple-span planar statically determinate trusses.

Another method used to determine the member forces in a truss is the use of a *graphic force polygon*, or *Maxwell diagram*, as developed by James Maxwell in 1864. Again, this procedure is discussed in detail in numerous engineering reference books and will not be repeated here.

Regardless of how carefully wood trusses are designed and fabricated, the final erection and bracing of a roof or floor truss system by the framing crew is critical. It is at this critical stage of construction that many significant design assumptions such as proper bracing are either fulfilled or ignored. If ignored, the consequences of inadequate bracing may be collapse of the structure.

The theory of bracing is to apply sufficient support at right angles to the plane of the truss to hold every truss member in the position assumed for it in design. The framer, however, is responsible for wood truss handling and for temporary and permanent bracing. The framer must assure that trusses are staged and maintained in alignment before and during installation and not in any way structurally damaged during installation. All roof trusses are built to exact specifications and are subject to the forces listed in each major model building code.

Temporary and permanent bracing may follow, but not necessarily be limited to the building designer's framing plan (which should be the source of the truss bracing and connection details). Standard framing procedure is that erection bracing is installed as each truss is placed in position.

Compression members are laterally restrained at specific locations or intervals. Superimposed dead or live loads act vertically, wind loads are applied normal to the plane of the top chord, and concentrated loads are applied at a point that erection bracing is installed as each truss is placed in position.

The design of wood trusses in accordance with standard building criteria assumes all of the following factors: Truss members are initially straight, uniform in cross section, and uniform in design properties. Trusses are plane structural components, installed vertically, braced to prevent lateral movement, and parallel to each other at the design spacing.

Truss members are pinned at points for determination of axial forces only.

There is a continuity of chord members at joints for determination of moment stresses. Compression members are laterally restrained at specific locations or intervals. Superimposed dead or live loads act vertically, wind loads are applied normal to the plane of the top chord, and concentrated loads are applied at a point.

In addition to the lateral bracing specified by the truss designer, the building designer will specify sufficient bracing at right angles to the plane of the truss to hold every truss member in the position assumed for it in design. The building designer (not the truss designer) will specify sufficient bracing and connections to withstand lateral loading of the entire structure.

Truss Installation

Always refer to the approved framing plans for the proper location of trusses. Do not simply install them "off the pile." Refer to truss elevation drawings for proper location of cantilever and interior bearings. If a truss is damaged before erection, obtain repair or replacement instructions from a qualified truss designer prior to setting the truss. Repairs made on the ground are usually better and always easier.

Never alter any part of any truss for any reason without the designer's authorization. Cutting, drilling, or notching any member without designer approval will destroy the structural integrity of the truss and void any expressed or implied manufacturer warranty. Truss girders are girders that carry other trusses or relatively large areas of roof or floor framing. Almost never will a common truss (or even a double common) serve as a girder. If there is any question about whether a girder is needed, consult the responsible architect. Double girders are commonplace. Even triple girders are sometimes required.

Make sure multiple member girders are properly laminated per engineering specifications. The nailing pattern must be per approved plans and specifications. In no case should nail spacing be more than 24 in o.c. Nails should not be less than 8d or greater than 16d. If more than one row of nailing is used, the rows must be offset at least $^1/_2$ in.

In some jurisdictions (check your local code), 14 gauge staples can be substituted for 8d nails when having a minimum penetration of 1 in into the main member.

Truss spacing should be in accordance with the truss design. An example of these specs appears in Table 7-5. Sometimes very small deviations from design spacing can create very big problems. Consult with the architect and truss manufacturer if you need to alter any spacing. Construction loads such as stacks of plywood, sheetrock, plies of roofing, gravel, or bricks should never be placed on trusses before they are properly braced.

Such loads should be kept small, located over walls (never at midspan), and distributed over the roof or floor within minutes after they are set in place. Make certain that temporary bracing is installed as the trusses are set.

Always consider the safety of others when stacking trusses. Never leave a framing job at night until all appropriate bracing is in place.

If trusses do not rest on all bearing walls properly, make certain that interior walls are not too high. If trusses are flat trusses, make certain they have been erected right side up. Correct with shims or by lowering high bearings. Trusses that do not meet interior bearing walls should be shimmed. Do not pull the trusses down to bearings. The prestressed truss will always tend to return to its original camber and in a few weeks sheetrockers will be repairing the joint

Table 7-5
Truss Specifications

Depth	Minimum Bearing Distance (in) Minimum Bearing at Maximum Shear (in)
10	$1^1/_2$
12	$1^5/_8$
14	2
16 to 20	$2^3/_4$

between the ceiling and wall. Guess who's going to get the bill?

Truss Terms and Definitions

Design Load. Top and bottom chord live and dead loads and allowable stress increase.

Frame (or Solid) Wall. Overhang length is the horizontal distance from the outside face of the plate or exterior wall surfacing to the bottom edge of the rafter (top chord).

Gable Ends. The specified angle of gable end required.

Nominal Span. The length of the bottom chord. Frame or exterior veneer over the frame wall span is the distance between the outside face of the wall or the outside face of the bearing plate.

Overhang Length. The horizontal distance from the end of the bottom chord to the bottom edge of the rafter (top chord).

Rafter End Cut. The proper angle of plumb cut, seat cut, or wild run (not trimmed).

Soffit Returns. The specified return dimensions for soffit framing.

Truss Count. Number of trusses required, factored at standard 24 in o.c.

Truss Roof Slope. Vertical rise in inches per foot of horizontal run.

Veneer over Frame Wall. Overhang length is the horizontal distance from the outside face of the plate or exterior wall surfacing to the bottom edge of the rafter (top chord), plus the width of the veneer, air space, and sheathing. If overhang length is not the same for both ends of the truss, the span includes the overhang for each end.

Web stiffeners are required at 18 in and above truss depths. Web stiffeners are field installed on each side of the web with 8d nailing spaced 3 in apart vertically (nails from either side counting). Nails must pass through the web and penetrate at least 1 in into the opposite stiffener.

Web stiffeners must be provided with blocking panels when specified. Web stiffeners may also be required at bearing points and where concentrated loads occur. Bridging should be installed as set forth in the local adopted building code.

Truss Configurations

The most common configurations of trusses are shown on the following pages in Figs. 7-13 and 7-14. These roof trusses serve as basically the same structural roof framing as do conventionally framed roofs.

Finishing the Eaves and Rakes

Before a roof can be shingled, the *eaves* (horizontal roof edges) and *rakes* (sloping roof edges) must be completed. When designing roof edge details, the designer should keep several objectives in mind. The siding, which is not installed until after the eaves and rake, should be easy to join to them. The edges of the roof shingles should be supported in such a way that water flowing over them will drip free of the trim and siding below.

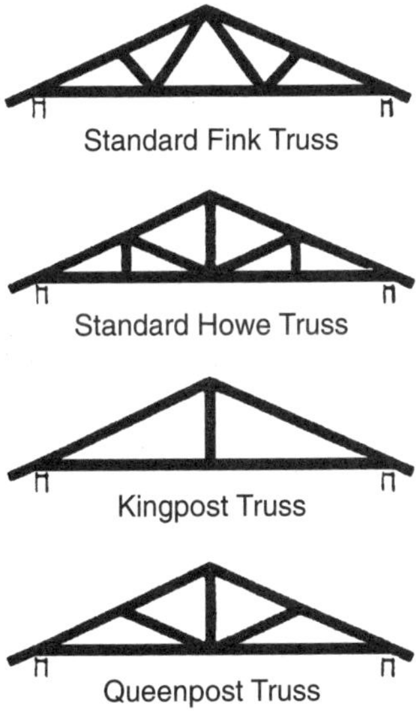

Figure 7-13 Common Trusses

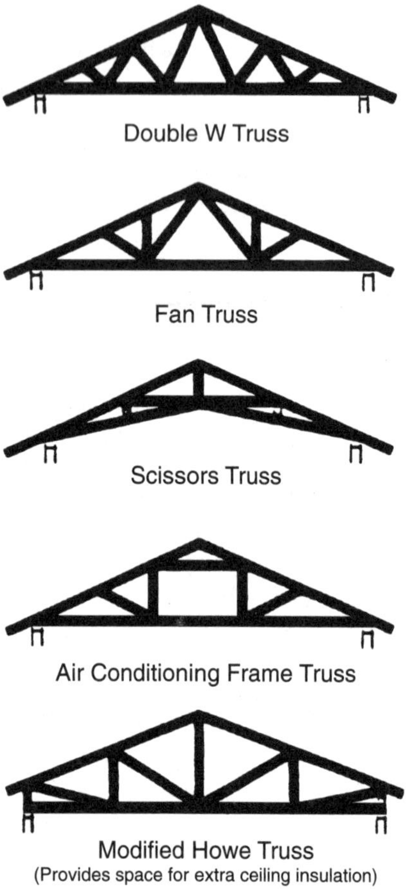

Figure 7-14 Common Trusses

The eaves must be ventilated to allow free circulation of air beneath the roof sheathing and provision must be made to drain rainwater and snow melt from the roof without damaging the structure below.

Roof Drainage

Gutters and *downspouts* (downspouts are also called *leaders*) are often installed on a pitched roof to remove rainwater and snow melt without wetting the walls or causing splashing or erosion on the ground below. Gutters built into the roof, once popular with architects, are difficult to construct sufficiently well to prevent water from entering the building when the gutters become clogged with debris. They have been superseded almost completely by external gutters of wood, aluminum, or plastic. These are sloped toward the points at which downspouts drain away the collected water.

Spacings and flow capacities of downspouts are determined by formulas found in the building codes. At the bottom of each downspout, a means must be provided for getting the water away from the building. A simple precast concrete splash block can spread the water and direct it away from the foundation, or a system of underground piping can collect the water from all the downspouts and conduct it to a storm sewer, a dry well (an underground seepage pit filled with broken rock), or a drainage ditch.

Many codes allow rainwater gutters to be omitted if the overhang of the roof is sufficient, thus avoiding the problems of clogging and ice buildup commonly associated with gutters. Typical minimum overhangs for gutterless buildings are 1 ft (300 mm) for single-story buildings, and twice this for two-story buildings. To prevent soil erosion and mud spatter, the drip line below the eave must be protected with a bed of crushed stone.

Ice Dams and Roof Ventilation

Snow on a pitched roof may be melted by heat escaping through the roof from the heated space below. At the eave, however, the shingles and gutter, which are not directly above heated space, can be very cold, and the snow melt can freeze and begin to build up layers of ice. Soon an *ice dam* forms, and water accumulates above it, seeping between the shingles, through the roof sheathing, and into the building, causing damage to walls and ceilings.

Ice dams can be largely eliminated by keeping the entire roof cold enough so that snow will not melt. This is done by ventilating the roof continuously with cold outside air through vents at the eave and ridge. In buildings with an attic, the attic itself is ventilated and kept as cold as possible, with insulation in the ceilings below to retain the heat in the building. Where there is no attic, the spaces between the rafters should be ventilated by means of air passages between the insulation and the roof sheathing.

Soffit vents create the required ventilation openings at the eave; these usually take the form of a continuous slot covered either with insect screening or with a perforated aluminum strip made especially for the purpose. Ventilation openings at the ridge can be either *gable vents* just below the roof line in each of the end walls of the attic or a *continuous ridge vent*, a screened cap that covers the ridge of the roof and draws air through gaps in the roof sheathing on either side of the ridge board. Building codes establish minimum area requirements for roof vents, based on the floor area of the building. Roof ventilation also keeps a structure cooler in summer by dissipating solar heat that has come through the shingles and roof sheathing.

Roof Shingling

Wood light frame buildings can be roofed with almost any standard roofing system. For reasons of economy and fire resistance, however, asphalt shingles are used in the majority of cases. These are applied either by the carpenters who build the frame or by a roofing subcontractor.

Wood shingle and shake roofs, clay and concrete tile roofs, and architectural sheet-metal roofs are also fairly common. Low-slope roofing, either built-up or a single-ply membrane, is easily applied to a light frame building.

Roof Ratios

Refer to the following pages regarding roof slope versus roof pitch. The slope of a roof is the ratio of the rise (2) to the run (1). The run is half the distance of the roof span (3). As shown in Figs. 7-15 and 7-16, the pitch indicates the incline of a roof as the ratio of the rise (1) to the total span (2) of the roof. It is expressed as a fraction. For example, if the rise is 6 ft and the span is 24 ft, the pitch is $^6/_{24}$ or $^1/_4$. As the roof gets steeper, the fraction becomes larger (e.g., $^1/_4$ is steeper than $^1/_5$, which is steeper than $^1/_6$).

Along with other factors of design, the roof pitch will determine the sheathing requirements. The sheathing can be either solid or open. Solid sheathing is made of plywood or boards laid side by side. When solid sheathing is made from boards, it is best to use tongue-and-groove (T&G) because of the interlocking edges creating a flush surface. If board sheathing less than 6 in wide is used, there will be a tendency of the sheathing to curl, and curled boards do not

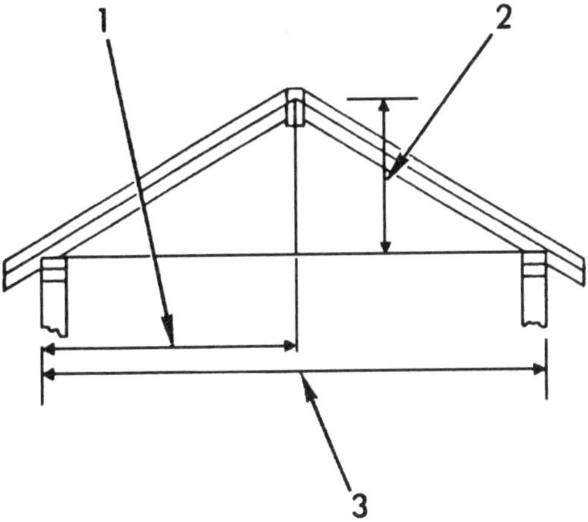

Figure 7-15 Roof Ratios

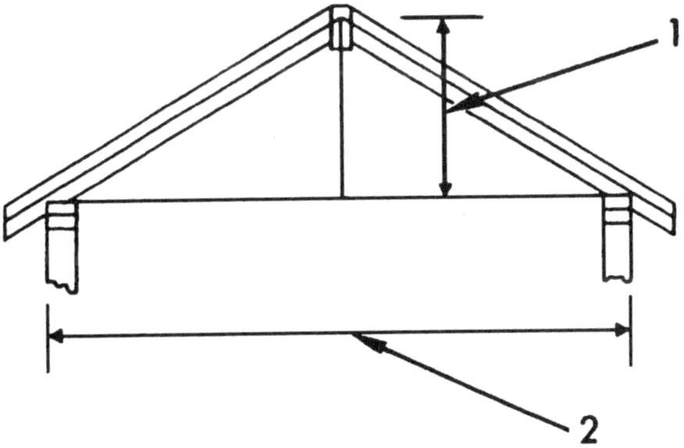

Figure 7-16 Roof Ratios

provide the necessary smooth surface that roofing surface materials require.

Open sheathing is made by laying narrow boards parallel to one another with a 2-in space between the boards. Wooden shingles and shakes are applied to open sheathing. Open sheathing allows the bottom surfaces of shingles and shakes to dry more rapidly than they would if applied to solid sheathing. Other types of roofing surface materials are attached to solid sheathing. (See Figs. 7-17 through 7-22.)

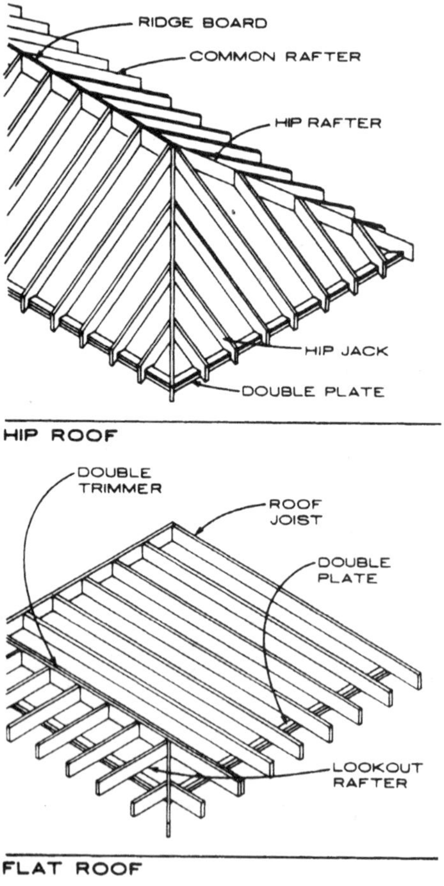

Figure 7-17 Roof Framing Details

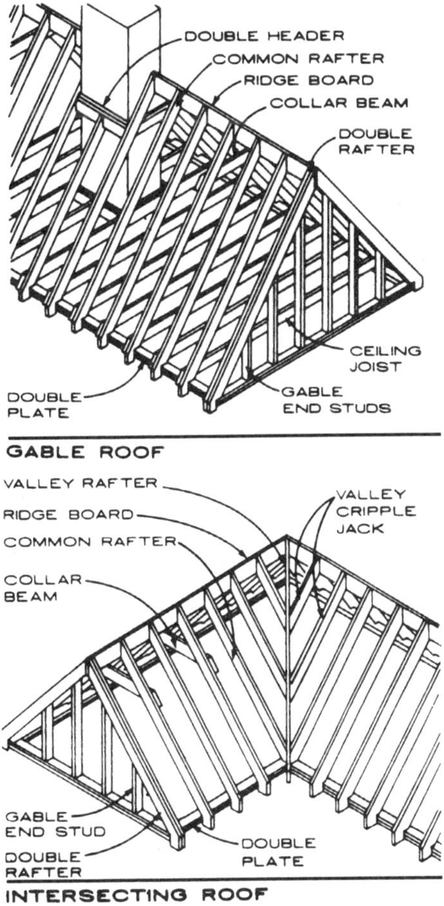

Figure 7-18 Roof Framing Details

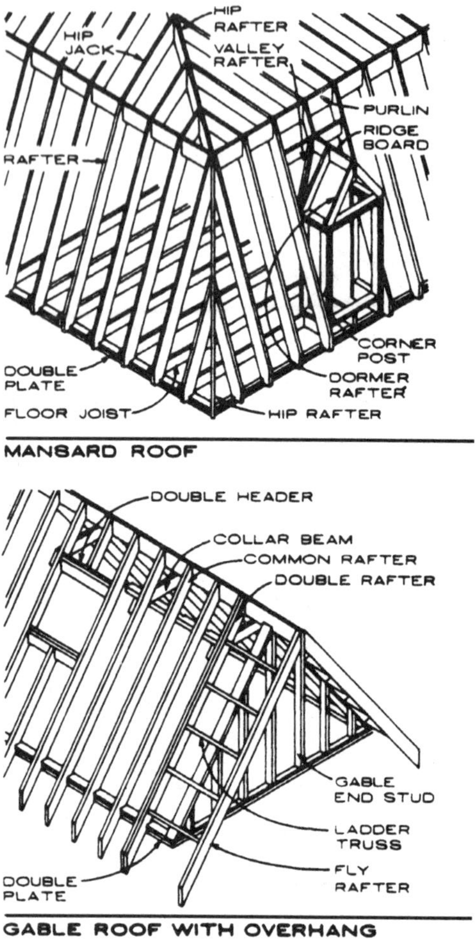

Figure 7-19 Roof Framing Details

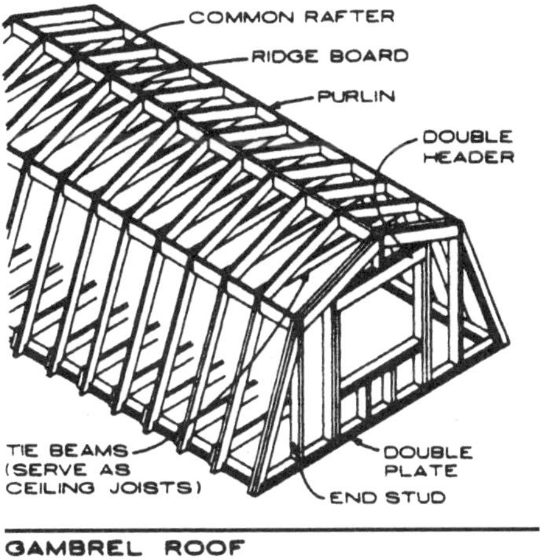

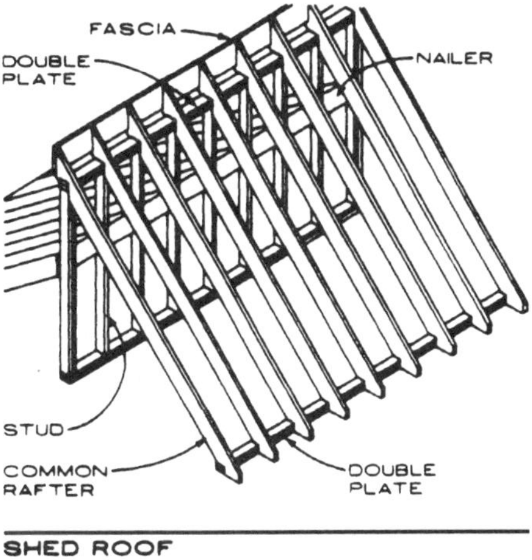

Figure 7-20 Roof Framing Details

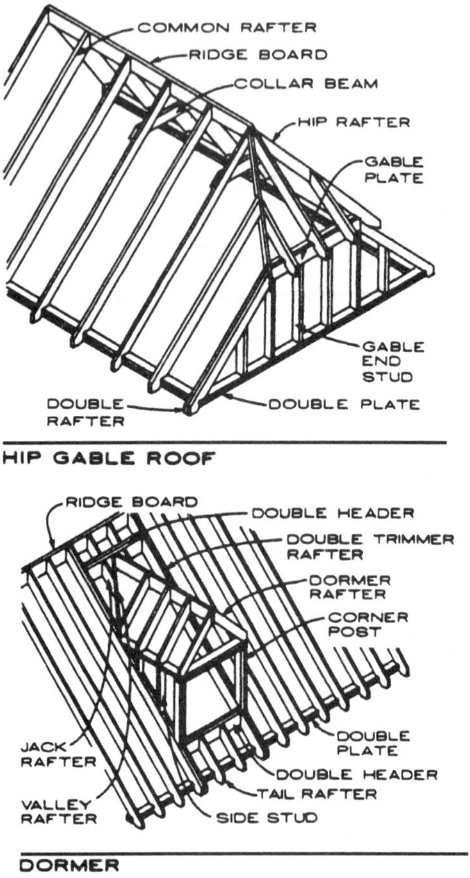

Figure 7-21 Roof Framing Details

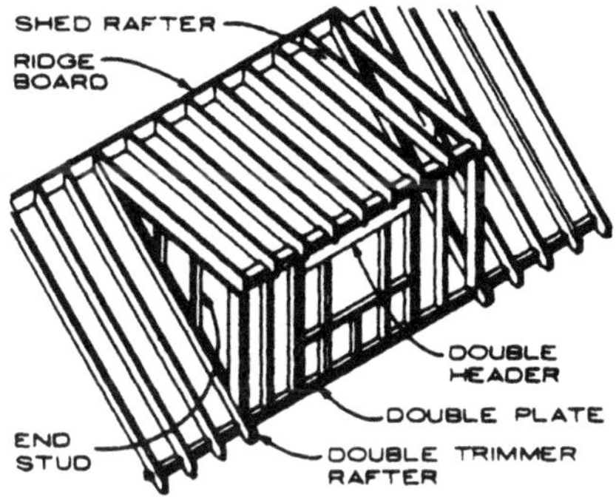

SMALL SHED DORMER

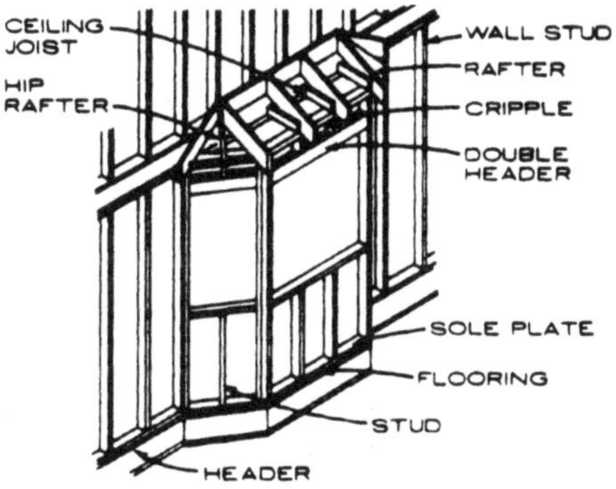

BAY WINDOW

Figure 7-22 Roof Framing Details

Notes

Chapter

8

Connectors and Holdowns

Connectors

A typical wood light framed building has many connections where shear forces are present. All of these connections are links in the shear load path. Just as with the uplift forces, these shear forces must be adequately resisted in order to minimize building damage.

No matter how heavy the gravity load, friction alone is not an effective way to create a load path. Remember that the heavier the gravity load, the greater the inertia and resulting shear force.

Two pieces of wood that are butted together can be easily pulled apart if not connected by some type of splice. Similarly, two pieces lying one on the other can be slid past each other if not properly connected. In wood framing, all connections need to have a nail, screw, or bolt connecting two pieces together.

Proper shear connections are created with combinations of fasteners, blocking, and hardware such as splice plates and framing anchors.

For nailed connections to be effective, the nails need to penetrate the wood a minimum depth. To accomplish this, the specified nail length must be used in all wood connections. It is also important to use the proper diameter nail (common vs. box) and avoid splitting the wood while nailing.

Even in the new technology structural framing applications, screws should not be randomly substituted for nails. While it is true that screws provide superior holding power in tension, they have reduced ductility. Ductility is necessary to prevent brittle fracture during cyclic loading. Always consult the architect or engineer and the local building official when considering this substitution.

A monolithic frame-to-foundation connector system to meet code specifications is engineered to act as a one-piece structural union between the two under seismic activity. These framing connector systems have also generated a number of options for architects and builders for a number of other common building problems. Often in remodel framing, it is impractical to use pressure-treated timber posts and dig new footings under older homes. The new framing connector systems are not so labor-intensive.

These connector systems work well in situations where corner caissons in foundations are subjected to seismic activity and keep shifting with the bedrock in which they are embedded—the perimeter footing is still monolithic, but the girder posts are off plumb. The framing connector system can be adjusted where necessary to pick up structural load points and replumb old ones. The exterior loadbearing walls are then connected to the perimeter footing by use of steel ties and grade A bolts or teco nails. The system can be further strengthened by lateral steel ties connected to each girder and welded or bolted to perimeter foundation ties.

The structural frame must also be anchored to the roof and all floors that provide lateral support for the wall. Structural tie connectors and fasteners act as anchorage that provides a direct connection between the structural frame and the loadbearing walls and the roof or floor construction.

Except for connections exempted by the local adopted building code, all parts of the structural frame that transmit seismic force must be interconnected to form a continuous path to the building's seismic resisting system. Any smaller portion of the frame must be tied to the remainder of the building with connectors having a strength capable of transmitting the seismic force F_p determined in accordance with code specs, but not less than one-third of the effective peak velocity-related acceleration A_v times the weight of the smaller portion W_c or 5 percent of the portion's weight, whichever is greater.

For a structural frame that is exempt from a full seismic analysis by code, the building's main wind force resisting system is deemed to be the seismic resisting system. A positive connection for resisting a horizontal force acting parallel to the framing member must be provided for each beam, girder, or truss to its support. The framing connection must have a minimum strength of 5 percent of the dead-plus live-load reaction.

In wall anchorage, the use of toe nailing or nails subject to withdrawal forces is not permitted. Wood ledgers cannot be subjected to cross grain bending or cross grain tension. The framing connectors must be capable of resisting a lateral seismic force F_p in accordance with code specs for bearing and nonbearing walls respectively, but not less than 1000 times the effective peak velocity-related acceleration A_v pounds per linear foot of wall. Walls must be designed to resist bending between anchors where the anchor spacing exceeds 4 ft (1219 mm).

Where openings occur in shear walls, diaphragms, or other plate-type elements, the edges of the openings must be built to transfer the stresses into the structure. The edge chord must extend into the body of the wall or diaphragm a distance sufficient to develop the stress of the chord member.

The design seismic forces are then applied separately in each of two orthogonal directions. Collector elements are provided that are capable of transferring the seismic forces originating in other portions of the building to the element providing the resistance to those forces.

The structural frame with connector system is designed to resist overturning effects caused by the seismic forces. At any story, the increment of overturning moment in the story under consideration must be distributed to the various vertical resisting elements in the same proportion as the distribution of the horizontal shears to those elements.

Strap tie connectors for $3^1/_2$-in or wider wood members are designed with 3-in nail spacing to minimize the chance of splitting the wood. Other specialized strap tie connectors are available when the I-joist flange is narrower than $3^1/_2$ in. These connectors come in a broad range of units suitable for wood-to-concrete or wood-to-block connection designed especially for wood I-joists and wood chord open web trusses.

The typical nail spacing is for 10d $\times$ $1^1/_2$-in nails on 3-in centers. This spacing minimizes the chance of splitting the supporting member. Light strap tie connectors are suitable in areas where gun-nailing is necessary through diaphragm decking. This type of connector is also used for ceiling support clips and furring stabilizer straps. It provides a 1-in separation between the furring channel and joist to allow for the use of insulation and the positive attachment of the furring channel to all joists.

These are designed to give an efficient sound barrier and a 1-hour UL listed fire rating. Furring stabilizer straps are used with ceiling support clip systems to prevent furring channel rotation during installation. Z-type clips are used for supporting 2 $\times$ 4 or 3 $\times$ 4 flat blocking between joists or trusses to support roof sheathing.

Tension bridging connectors use two 10d $\times$ $1^1/_2$-in nails per end that work only in tension, so they must be installed in crossed pairs. Install bridging fairly tightly as loose installation may allow floor movement. Avoid contact between metal bridging, which may cause squeaks. For installation at a bearing location (for the connection to the bearing plate instead of joist to joist), use a strap length one size smaller than shown in the load tables.

Industry standard strap tie framing connectors for column caps and bases are designed for use with composite beams and columns. Use all manufacturer's specified fasteners and use the preset tabs to temporarily secure members while drilling and bolting. Column caps only may be ordered for field welding to pipe or other columns. Any two column caps of this type can be specified for back-to-back welding to create a crossbeam connector. For loads, use the code table; however, standard specs are that the load be no greater than the lesser element employed.

Connectors for composite wood products now must provide for opposing connectors when used with Douglas fir larch, Southern Pine, and Spruce-Pine-Fir. In particular, the U/HU hanger series (which are U-shaped holdown hangers for header-to-joist connections) has an extensive product listing. All connector loads reflect the new standards adopted by the International Conference of Building Officials (ICBO), specified in the 1991 National Design Specifications. The "Minimum Nailing" and "HU Maximum Nailing" options allow installation for two different load requirements using different nailing configurations.

Strap ties and purlin anchors extend the current strap-tie series with "Z4" and "Z6" clips that support flat blocking between trusses that are load-rated.

Holdowns

Tall and narrow shear walls have a tendency to overturn owing to uplift forces created when an earthquake (lateral) force pushes the top of the wall. When the uplift forces exceed the given weight on or in the wall, additional resistance must be applied. This is done with the installation of metal hardware typically identified as a holdown. The engineer or architect will specify the installation and locations of holdowns when anticipated uplift forces exceed gravity loads on the walls.

Holdowns transfer uplift forces from the end of the wall through a floor to a wall or foundation below. When an earthquake shakes a shear wall back and forth, the shear wall will experience uplift forces on both ends, one end at a time. It is for this reason that holdowns are typically required at both ends of a shearwall and are connected to the end stud or post of the shear wall. The added post is better able to resist the tension load from uplift and the simultaneous compression load at the opposite end of the shear wall.

Holdowns that connect a shear wall to the foundation are bolted, screwed, or nailed to the end stud or added post. Retrofit holdowns are anchored to the footings using a threaded rod that is connected to the holdown. The other end of the rod is adhesive set into a hole that is drilled into the foundation concrete. This rod transfers the uplift force from the wall down into the foundation. This minimizes the wall's tendency to overturn.

To better understand the importance to the structural frame of properly installed holdowns, consider a tall and narrow shear wall that is 8 ft tall and 2 ft 3 in wide (height-to-width ratio of $3^1/_2$ to 1). By applying an earthquake-induced horizontal load at the top of the wall, the wall will begin to overturn. For every inch that the bottom left corner of the wall goes up, the upper right corner will move laterally a distance of $3^1/_2$ in.

Cyclic testing of wood shear walls has shown that lateral movement of as little as 2 in can cause loss of vertical support and therefore cause structural frame collapse. Accounting for the height-to-width ratio of $3^1/_2$ to 1, this means that if a holdown at the bottom left corner of a shear wall slips as little as $^5/_8$ in, the entire building could suffer significant damage and possible collapse.

Holdown hanger connectors are individually selected based on the connectors' steel specifications, including shear strength, thickness, formability, finish, and weldability. Allowable loads are based on use with Group II lumber species Douglas fir-larch, Southern Pine (based on 565 psi), or Group III lumber species Spruce-Pine-Fir (based on 425 psi), under continuously dry conditions, depending on the table used.

Allowable loads for other species must be adjusted according to the code. Gauge may vary from that specified depending on the manufacturing process used, however, any variance from design engineering specs must be approved in advance by the local building authority.

U and W type holdown hangers normally have single stirrups; GLT holdown hangers for sloped seat installations are all assumed to be nonbacked and manufactured to accommodate the quantity of fasteners needed to achieve the comparable standard model's table load. All references to bolts or mechanical fasteners are for structural quality through bolts equal to or better than ASTM Standard A307.

If ordered for a fully backed installation, sloped down only holdown hangers require the number of nails in the table. Unless otherwise specified, use sloped seat holdown hangers when. slopes exceed $^1/_2$:12 for bearing length of $2^1/_2$ in or less; $^3/_8$:12 for bearing length between $2^1/_2$ and $3^1/_2$ in; and $^1/_4$:12 for longer bearing lengths.

Slope conversions are shown in Table 8-1.

Welding galvanized steel will produce toxic fumes; follow proper welding procedures and safety precautions. Die-formed holdown hangers are tempered and cannot legally be modified by welding. However, most manufacturers offer models that will normally accommodate a skew of up to 5°. Heating, reforming, or bending steel in the field may cause fractures at the bend and is not allowed by code.

Fractured steel will not carry design load and must be replaced. Connector hanger allowable loads are based on the lowest of the following: the test ultimate divided by 3; the load producing a $^1/_8$-in deflection; the fastener values; and the seat

Table 8-1
Slope Conversions

Rise/Run	Slope
$^1/_{12}$	5°
$^2/_{12}$	10°
$^3/_{12}$	14°
$^4/_{12}$	18°
$^5/_{12}$	23°
$^6/_{12}$	27°
$^7/_{12}$	30°
$^8/_{12}$	34°
$^9/_{12}$	37°
$^{10}/_{12}$	40°
$^{11}/_{12}$	42°
$^{12}/_{12}$	45°

bearing value plus the allowable carried member fastener rating.

Holdown connectors other than holdown hangers are load-rated using a similar method. Manufactured connectors are tested to failure using industry standard procedures. Allowable loads for more than one direction for a single connection cannot be added together. A design load that can be divided into components in the directions given must be less than or equal to the lowest allowable table load listed.

Some combinations of hanger options have not yet been evaluated, so always check with local code regulations. Wood shear is not considered in given load factors, so the framer must reduce allowable loads when wood shear is limiting.

For headers, the top flange may be offset left or right for placement at the end of the header. The allowable load is 0.50 of the table loads. For skewed and offset top flange holdown hangers, the maximum allowable load is 0.50 of the table load, or 2600 lb, whichever is lower. Verify that the header can take the required fasteners specified in the table and use all of the specified manufacturer's fasteners. Nailers and ledgers must be a minimum of 4x lumber to guarantee the load values given in the tables. Thinner lumber or laminated veneer lumber used as a nailer must be evaluated by the building designer.

For joist systems, multiple joists should be adequately connected together to act as one unit. Holdown hangers may support joists from twisting (torsion) unless the joist is laterally restrained by other means. Joist hanger side flanges should be at least 60 percent of joist depth and support the joist top flange. Joists should be supported by bearing on the bottom flange, otherwise the joist's shear capacity may be reduced and the web section could pull away from the bottom flange (unless designed otherwise).

Lack of a web stiffener installed on the joist prior to hanger installation will result in torsion unless restrained by other means. Solid sawn web stiffeners or backer blocks may split if a row of closely spaced nails goes along the wood grain. Nail patterns should be staggered so as to not split joists, backer blocks, or web stiffeners. Roof loads are 125 percent of normal floor loads unless limited by other criteria. Floor loads may be adjusted for other durations according to the code, provided they do not exceed those for roof loads. Code regulations on uplift loads have been increased 33 percent for wind or earthquake loading with no further increase allowed. Reduce by 33 percent for normal loading such as in cantilever construction.

For I-joist to I-joist connections, the nails should extend through backer block and web. IUT-type holdown hangers use a bend tab that nails vertically into the I-joist's bottom flange when web stiffeners are not used, or directly into the web stiffener. This constrains the member, helping to reduce squeaks resulting from joist movement.

The top flange hanger's supporting wood I-joist may require bottom flange restraining straps, blocking, or directly applied ceiling systems to prevent rotation at the hanger location. When supporting I-joists from one another, install backer blocks against the bottom of the top flange of the carrying I-joist. This prevents cross-grain bending and rotation of the top flange.

Built-up lumber (multiple members) must be fastened together to act as one unit to resist the applied load. GLTV- and HGLTV-type holdown hangers are designed for use with laminated veneer lumber (LVL) or parallel strand lumber (PSL) headers, and may take heavy loads. For slope only, the maximum allowable load is 6500 lb for the GLTV hanger and 9165 lb for the HGLTV hanger. For skew only, the maximum allowable load is 6550 lb for the GLTV hanger, and 10,670 lb for the HGLTV hanger. The HGLTV hanger may have $^1/_4$ in deflection at full loading.

The top flange nails are sized and specifically located to prevent degradation of the header due to splitting of laminations. A top flange may be sloped down left or down right to 30° with or without a sloped and/or skewed seat. Reduce allowable table loads using straight-line interpolation. For hanger heights exceeding the joist height, the allowable load is 50 percent of the table load. Consult local code regulations for hanger maximum loads, dimensions for installation, allowable loads, and fastener requirements.

The three major model national building codes (ICBO, BOCA, and SBCCI) all have similar specifications concerning the corrosion resistance factors of loadbearing connectors. These specifications are based on the fact that connectors are steel and will corrode and lose load-carrying capacity if exposed to ocean salt air, corrosive fire-retardant chemicals, fertilizers, or other substances that adversely affect steel. In corrosive environments, especially for connectors that are not exposed to rain or periodic washing, deterioration will occur at a faster rate. The service life of galvanized connectors in a temperate marine environment can be greatly improved by proper maintenance. Fasteners of comparable material should be used to install the product.

Currently, steel is provided this corrosion resistance by three methods. The first is hot-dip galvanized, in which the products are hot-dipped after fabrication. The coating weight increases with material thickness. Hot-dip galvanizing is available for many products, and these holdown hangers require galvanized fasteners. The second is Z-max (G185) triple zinc coating. This produces galvanized products that provide three times the zinc of standard galvanized products. Triple coat Z-max meets all load specifications and all published building code evaluation reports.

These also require galvanized fasteners. The third (and best) is by using SST300 stainless steel holdowns. These holdowns are made from Type 316 stainless steel and provide much greater durability against corrosion than galvanization. Stainless steel nails only should be used with stainless steel connectors.

Holdown hangers are heavy-duty hangers designed for multistory framing and other structures demanding increased strength, longevity, and safety factors. New designs allow for high loads with common nails. Web stiffeners are required with all I-joists using this hanger style. Holdown hangers with a greater than 15° skew may have all joist nailing on the outside.

Holdown hangers may be skewed to a maximum of 67° and/or sloped to a maximum of 45°. For slope only or skew only, the allowable downloads are 100 percent of the table loads. For skews greater than 15°, uplift loads are 75 percent of the table loads. For combined slopes and skews, the maximum allowable download is 80 percent of the table maximum load. Nonskewed holdown framing hangers are capable of carrying the design load when the seat slope is within 2° of the actual joist slope. Local crushing of the bottom chord or excessive deflection may be limiting; check with the joist manufacturer for specific limitations on wood bearing of this type. Standard industry tolerances for nonskewed holdown hangers allow for small adjustments of the joist for skew angles in the field. The nonskewed models will normally accommodate a skew up to 2.5°.

Double shear holdown hangers use double shear nails, which have the advantage or greater connector strength. Double shear nailing requires fewer nails, allows faster installation, and gives more strength than comparable holdown hangers. An added advantage is that common nails can be used for all connections instead of more expensive joist hanger nails (tecos). Face nails attach to both upper and lower top flanges, helping them to act as one unit.

Typical installation is to install top nails to face the pan nails in the A flange to the outside wall top plate, then seat the rafter with a hammer, adjusting B flange to the required pitch. Install the B flange nails in the oblong nail holes, locking in the pitch. Lastly, install rafter pan nails. Code specs are that a minimum of $2^1/_2$ in of top flange material is available for minimum nailing configuration. For nonhip roof connections, special holdown hangers provide a top flange bearing mount instead of a face mount, which is ideal for top plate nailing applications. Consult local code regulations for hanger maximum loads, dimensions for installation, allowable loads, and fastener requirements.

Adjustable truss holdown hangers feature double-shear nailing for added strength and safety. Face-mounted holdown hangers provide support for combined loads up to 3100 lb. Matched load capacity holdown hangers handle up to 4200 lb with no bolts required for installation. These provide support when compression perpendicular grain limits truss or girder-to-plate conditions.

As per building code requirements, they replace nail-on scabs for load transfer. Single piece, nonwelded holdown connectors are engineered to carry multiple truss members such as terminal scissors applications. They are also available in one-piece welded units skewed 45 to 67°. Other applications are sloped by design up to 30 percent with only a 10 percent load reduction.

Common Holdown Installation Errors

Holdowns should be installed as specified on the plans and by following the manufacturer's installation instructions. Substitutions of holdown hardware require the prior approval of the building department and the design professional (engineer or architect). An errant substitution may not provide the same resistance to uplift as originally intended. When new holdown posts are specified, such as a 4 × 4 or 4 × 6, you should not substitute the specified post with a doubled stud. The allowable strength of a holdown is partly based on the cross-sectional area of the end stud. The strength of the holdown is generally reduced by attaching it to two 2 × 4 studs (3 in total thickness) instead of the specified solid post ($3^1/_2$ in total thickness).

When nailing on straps used as holdowns, the minimum required edge distance must be maintained. The minimum required edge distance depends on the size of nail used. Nails should be installed no closer to the edge of the wood member than one-half the minimum required embedment depth of the nail (see Table 23-G of the Uniform Building Code). For example, if the nail's minimum embedment depth is $1^3/_4$ in, the nail should be installed no closer than $^7/_8$ in from the edge of the wood member.

Shear wall end studs or posts should normally be one full piece for the height of the wall. This is necessary because the end stud or post must be capable of resisting tension due to the uplift forces. Otherwise, additional connections will be needed inside the wall to transfer the uplift forces. When plans specify the installation of new posts in an existing or new wall, the posts should be installed at the ends of the wall. As you now know, tension or uplift loads occur at the ends of a shear wall. Unless otherwise approved by the engineer or architect, holdowns should be located as close to the ends of the shear wall as practical.

With respect to the holdown location on the post, the bolts attaching the holdown to the post should be located per the manufacturer's installation instructions. In general, the first bolt should be at least seven bolt diameters from the bottom of the post. This minimizes the chances of the bolt's ripping right through the end of the post during an earthquake.

The post bolt holes must not be oversized. If the hole is bigger than the bolt, the bolt will move before it engages the wood and will be much more likely to split the post. The building codes allow the hole in the end stud to be a maximum $^1/_{16}$ in oversized. The proper drill bit size must be used so that the hole is no larger than $^1/_{16}$ in bigger than the bolt diameter.

In addition, washers should always be used where the bolt head or nut bears directly on wood. This typically occurs on the post side that is opposite the holdown. Square plate washers should be used instead of round malleable washers. This will help better prevent the bolt pulling through the end stud during an earthquake. When through bolts are specified, lag screws should not be used. A lag screw, even if it is the same diameter, will not have the same strength as a through bolt.

Post bolts should never be countersunk into the post unless the installed post is at least one size larger than that specified. For instance, if the plan calls for a 4 × 4 post and it becomes necessary to countersink the bolt head, a 4 × 6 post should be used. This allows for an additional 2 in of post thickness in which to countersink the bolt head.

Holdown rods anchored into the foundation should be embedded as required by the manufacturer or structural design. The rods should be embedded only into sound concrete or reinforced masonry. Unlike a typical anchor bolt, the holdown rod must resist uplift forces. The loads imposed on a holdown rod are generally greater than those imposed on a typical shear wall anchor bolt.

Holdown anchors that are not installed to the proper depth and at the minimum edge distance will not perform as intended during an earthquake. Wedge- or expansion-type anchors should not be substituted when chemical (adhesive) anchors are specified. For holdowns, these anchors do not generally perform as well as adhesive anchors during the cyclic uplift loading caused by earthquakes.

The nut must be tightened securely on the holdown rod. If not properly tightened, the wall will begin to uplift during an earthquake before it engages the holdown. This movement can split both the sill plate and the holdown post. When the nut is not properly tightened, shear wall overturning will cause damage to the structural frame.

Retrofit Framing

When evaluating an existing building for possible retrofit, it is important to identify nonstandard framing. Balloon framing is one type of nonstandard framing where the intermediate floor framing joists are face nailed directly to the studs.

Because the studs are continuous from top to bottom of the building, this creates a load path discontinuity between the floor diaphragm and shear wall. When encountering this or any other type of nonstandard framing, an architect or engineer should be consulted.

There are many breaks in the continuity of a conventional platform-framed building, particularly if it rests on cripple walls. Shear forces created by the earthquake must be effectively transferred through these breaks in order for the load path to be complete. The following elements must be connected to each other to develop this complete path.

- Roof and ceiling must be attached to the top plate(s)
- Top plate must be attached to the sheathing
- Sheathing must be attached to the sole plate
- Sole plate must be attached to the floor framing
- Floor framing must be attached to the top plate(s)

- Top plate(s) must be attached to the sheathing
- Sheathing must be attached to the foundation sill plate
- Foundation sill plate must be attached to the foundation

When these elements are properly connected, they form an unbroken load path from the roof and floor diaphragms to the ground. (See the section on load path describing and showing the 44 elements in this shear load path.)

In retrofit framing, structural wall panel edges should be centered on all framing members with a $^1/_8$-in space between them. The spacing allows for future panel expansion and contraction from moisture and temperature changes. If panels are not properly spaced, expansion will cause buckling. Buckling reduces shear wall performance and can damage wall finishes. Panel edges should also have blocking to provide proper fastener edge distance in both the lumber and the sheathing.

The greater the fastener edge distances in the lumber and sheathing, the better the shear wall will perform. Fasteners installed too close to the edge of the sheathing will cause the blocking or framing member to have little or no resistance value.

Unlike unblocked floor and roof diaphragms, shear wall strength values in the codes do not change with orientation of the panel. Either vertical or horizontal placement of the individual panel is acceptable, however, most panels are installed vertically to avoid the added expense of blocking. See Fig. 8-1. Sheathing for shear walls may be placed on either the interior or exterior face of the wall framing.

Wood structural panel sheathing is usually placed on the exterior for new construction and on the interior for retrofit work. For new construction, work of other trades, like insulation and electrical, is generally easier when the interior is accessible. For retrofit work, the interior wall surfaces are usually cheaper to replace. The shear wall will have the same strength with the sheathing on either side of the wall.

Although the code does not currently prescribe minimum dimensions for individual panels on shear walls, a minimum width of 24 in is recommended to prevent any one

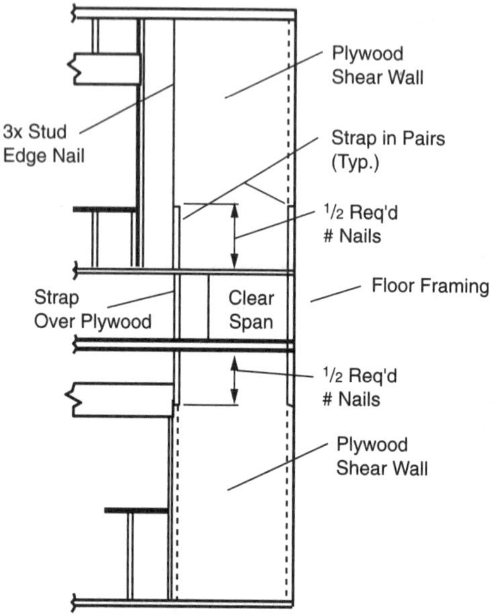

Figure 8-1 Vertical Shear Panels

local defect from affecting total performance. This criterion is used for floor loads and is recommended for shear walls. To follow the code recommendations of 2:1 aspect ratios in seismic zones 3 and 4, contractors should always install at least one full 4 × 8 ft panel on any standard 8-ft-high shear wall. When the length of the shear wall exceeds 4 ft, the end panels should be at least 24 in.

Retrofit work is usually done from one side of the wall. Panels are held in place with a few nails until volume nailing is done later. Unlike new construction, the number of fasteners that do not properly connect to the framing members or blocking cannot easily be determined after the sheathing is in place.

To reduce the chance of improper fastening, the framer should mark the center of all framing members, blocking, and holdown end studs during sheathing installation. After all panels are in place, a chalk line can be snapped on the panels, which will make proper volume fastening easy and accurate.

When retrofit work replaces the existing wall finishes, any fire-resistive construction must be maintained. One-hour fire-resistive and sound-rated construction is required at walls separating dwelling units in the same building. One-hour fire-resistive construction is also required throughout all three-story apartments and two-story apartments that have more than 3000 sq ft on the second floor.

Some light wood framed apartments may also use 2-hour fire-rated walls to get more allowable floor area. When these conditions exist, all required components of the wall assembly must be replaced. The addition of plywood or OSB panels will not affect the fire rating when attached directly to the framing and covered with the appropriate fire-resistive materials.

Another critical condition framers often encounter in retrofit work with older homes is a lack of building paper under their horizontal siding around the crawl space. This permits water to enter the cripple wall stud cavity. Circular ventilation holes should be cut in the sheathing when new sheathing is placed on the inside face of the cripple wall. Prescriptive standards specify the size and spacing of the ventilation holes. The holes will permit water vapor to escape and will permit inspectors to observe and test anchor bolts. Screens should be installed over the holes to prevent animals from nesting in the newly sheathed stud cavities.

New sheathing should not block existing vents. When sheathing must be replaced where a ventilation or other opening is located, cut an opening for the vent, install blocking around the vent opening, and extend the length of the shear wall the width of the vent opening to the nearest stud. The sheathing will be checked in the frame inspection, which is done after the exterior and interior walls are framed; fire stops, freeze blocks, and lateral bracing are installed; electrical and plumbing are roughed in; and all vents are installed.

Modern connectors and holdowns are illustrated in detail on the following pages in Figs. 8-2 through 8-19, and appear courtesy of Simpson Strong-Tie Company, the industry leaders in structural framing connectors and holdowns.

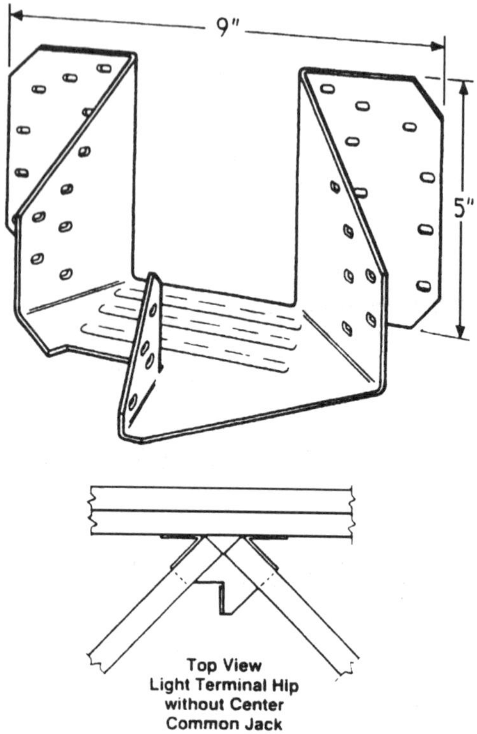

Figure 8-2 Modern Framing Connectors

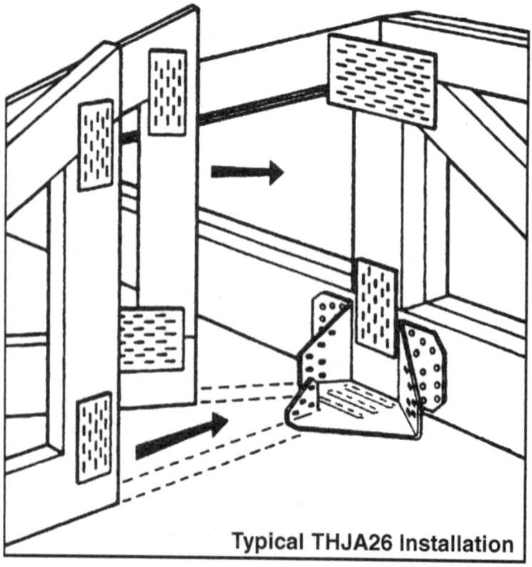

Figure 8-3 Modern Framing Connectors

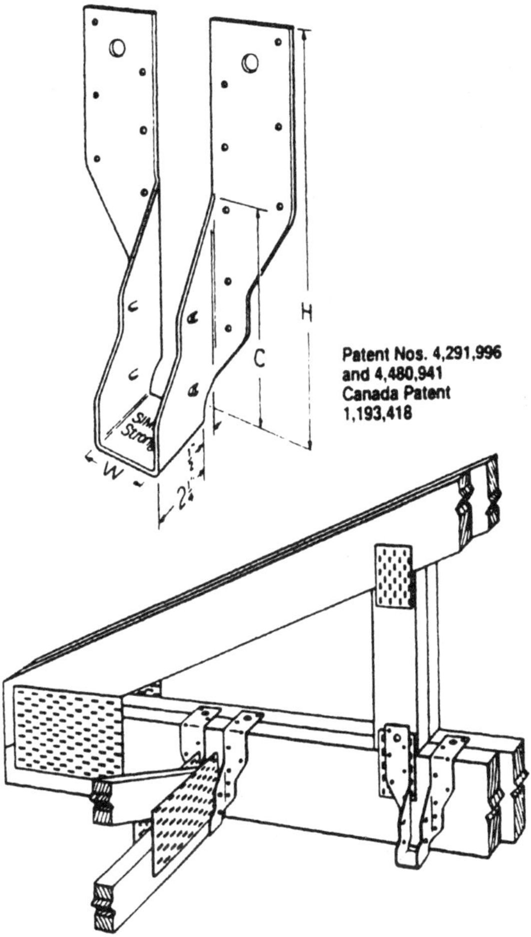

Figure 8-4 Modern Framing Connectors

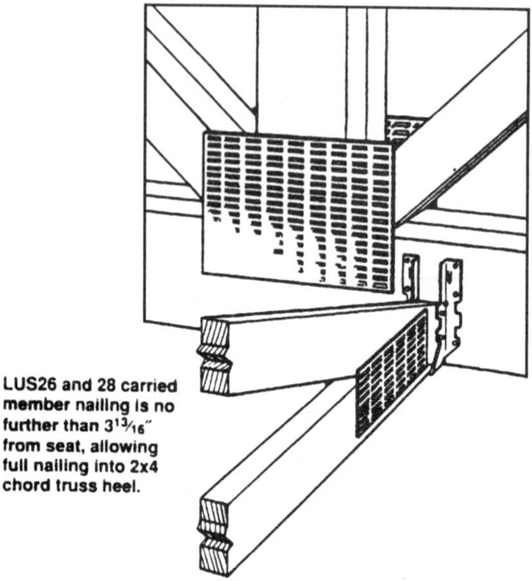

Figure 8-5 Modern Framing Connectors

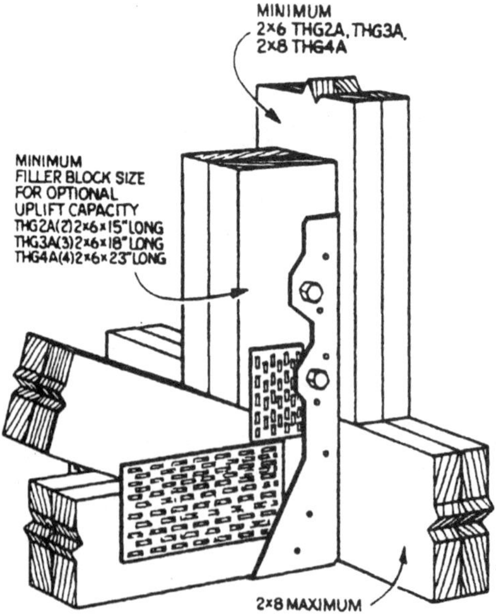

Figure 8-6 Modern Framing Connectors

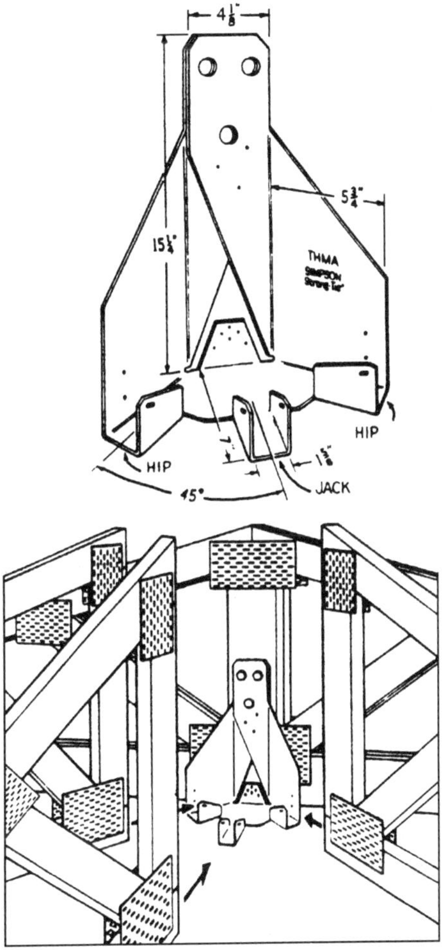

Figure 8-7 Modern Framing Connectors

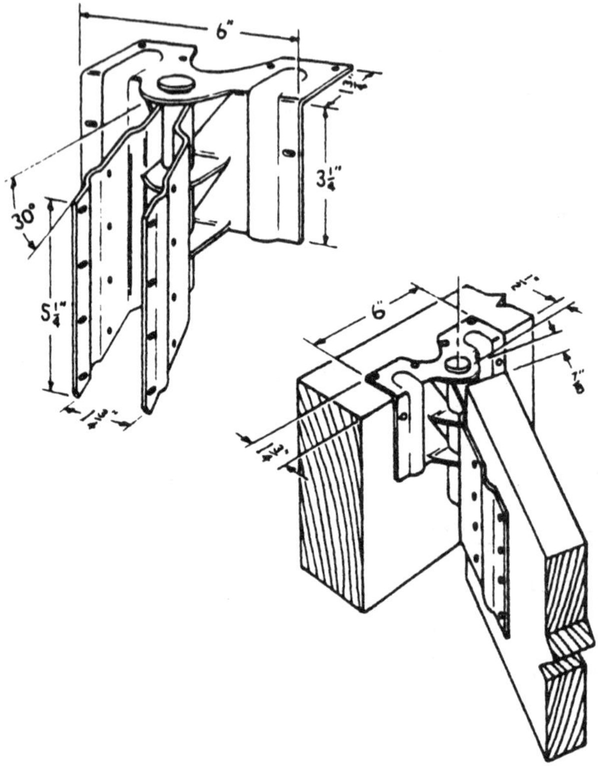

Figure 8-8 Modern Framing Connectors

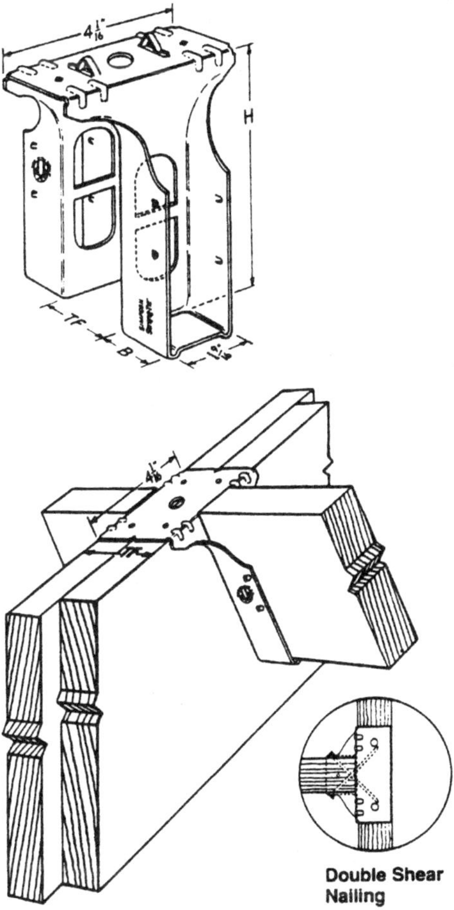

Figure 8-9 Modern Framing Connectors

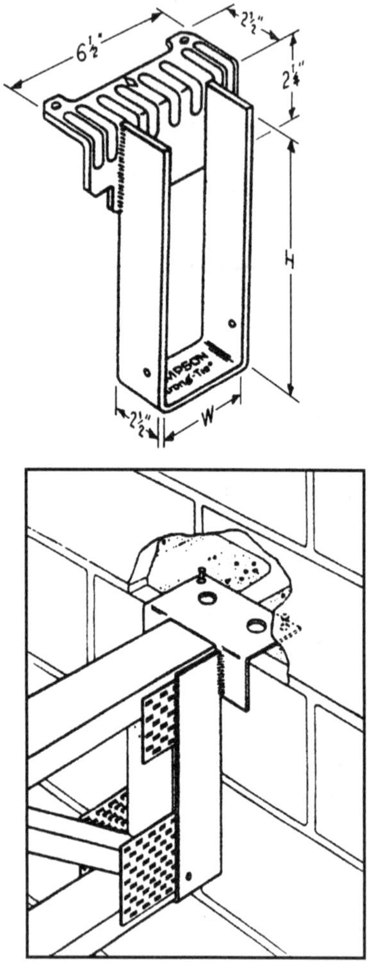

Figure 8-10 Modern Framing Connectors

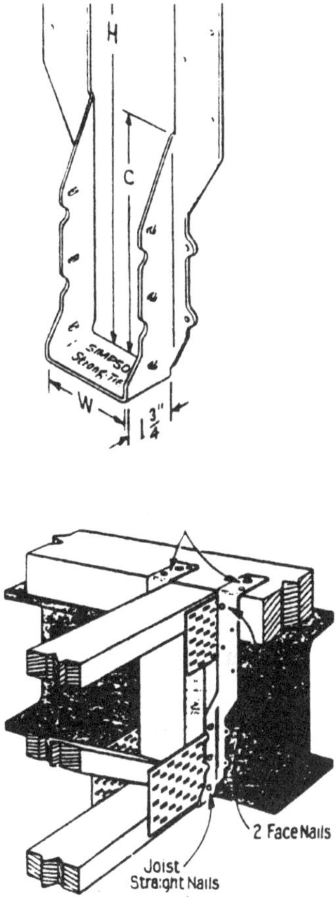

Figure 8-11 Modern Framing Connectors

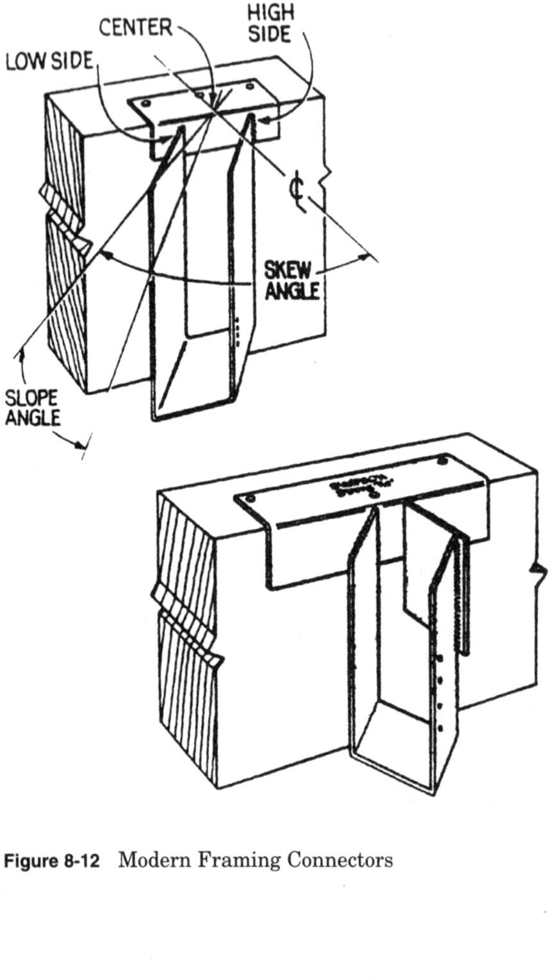

Figure 8-12 Modern Framing Connectors

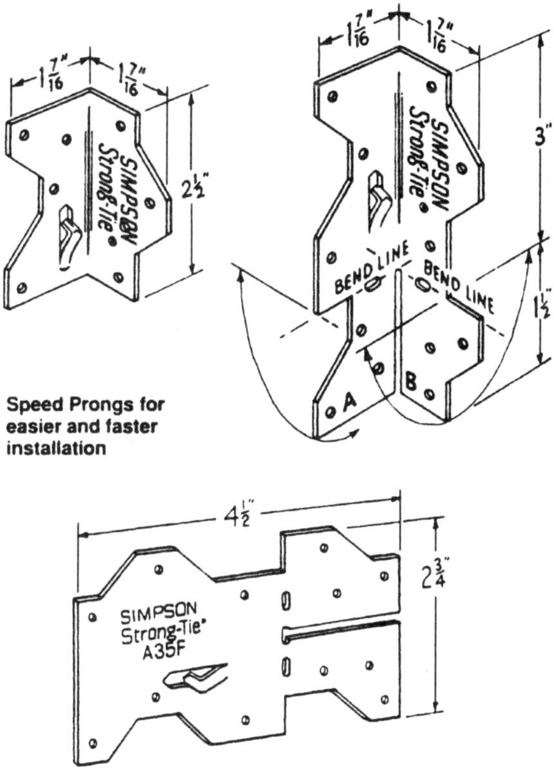

Figure 8-13 Modern Framing Connectors

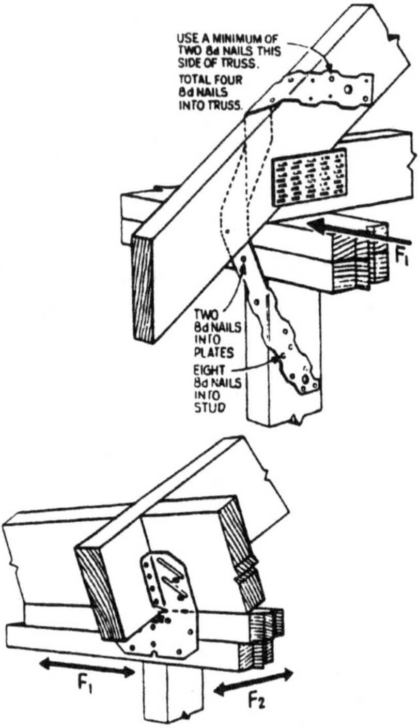

Figure 8-14 Modern Framing Connectors

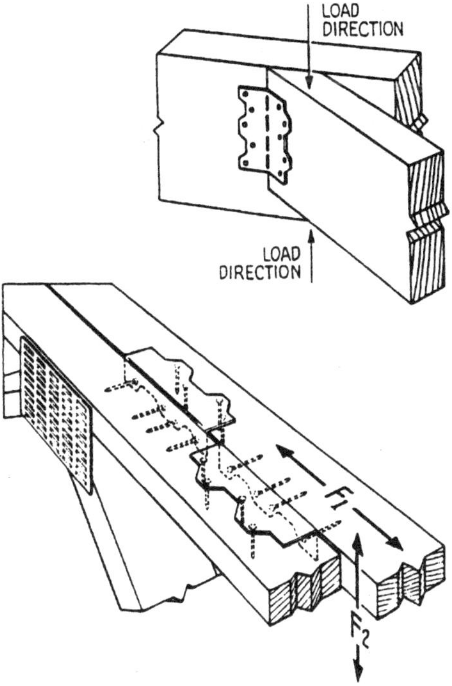

Figure 8-15 Modern Framing Connectors

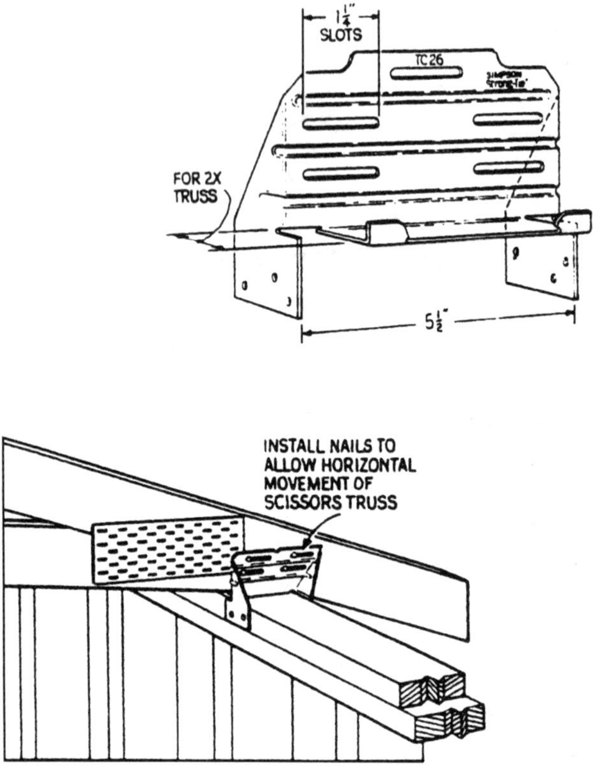

Figure 8-16 Modern Framing Connectors

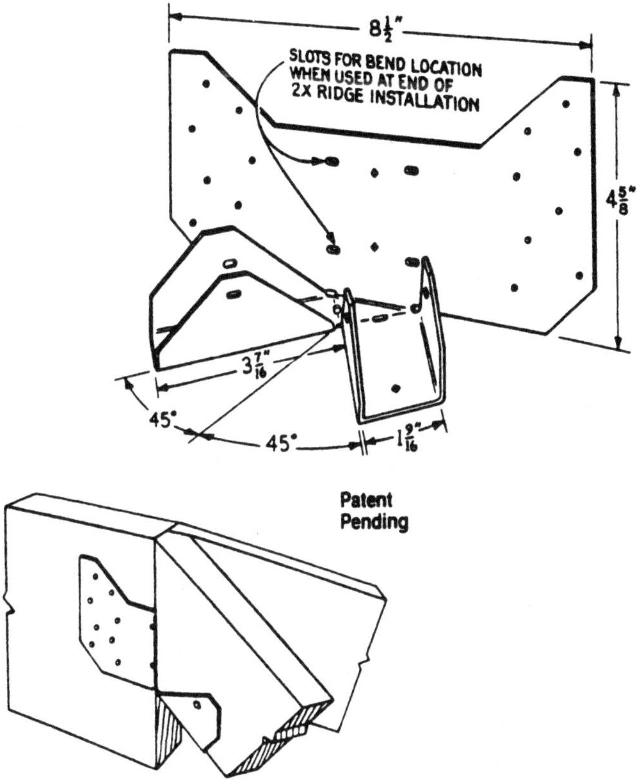

Figure 8-17 Modern Framing Connectors

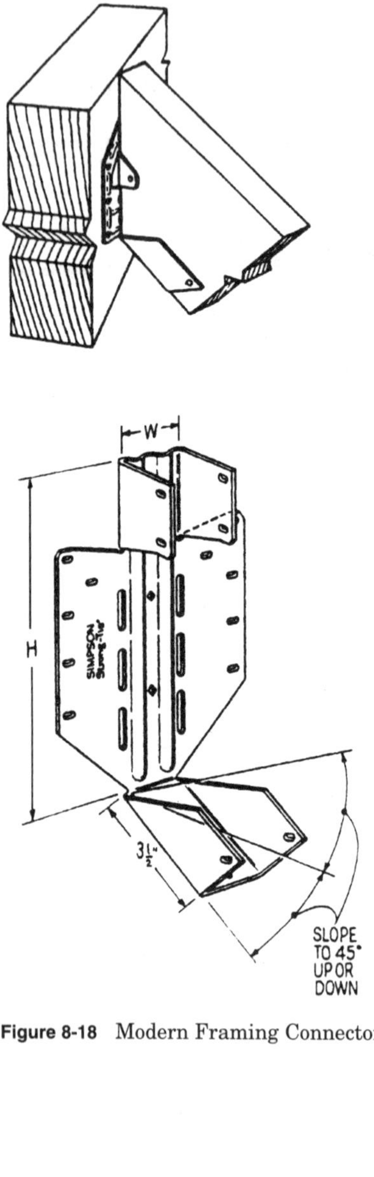

Figure 8-18 Modern Framing Connectors

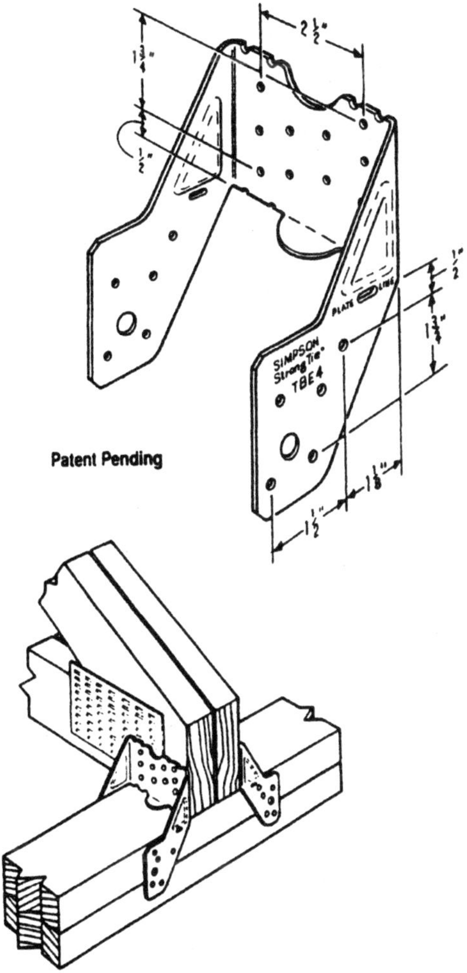

Figure 8-19 Modern Framing Connectors

Notes

Appendix

Framing Terms and Definitions

Acceleration The rate of change (increase or decrease) in velocity. As seismic waves travel through the earth, the ground moves backward and forward, changing its velocity; acceleration is related to velocity and displacement.

Adhesive Anchor An assembly consisting of a threaded rod, washer, nut, and chemical adhesive for connections to existing concrete or brick elements. Chemical adhesives may be epoxy, esters, or acrylics.

Anchor Bolt A cast-in place bolt used to connect the foundation sill to the foundation.

Braced Frame An essentially vertical truss, or its equivalent, of the concentric or eccentric type that is provided in a bearing wall, building frame, or dual system to resist seismic forces.

Building Frame A structural framing system with an essentially complete space frame providing support for vertical loads. Seismic force resistance is provided by shear walls or braced frames.

Compression When a wood member resists a pushing force along its axis on each end toward its center, typical of columns, posts, and holdown studs.

Concentrically Braced Frame (EBF) A diagonally braced frame in which at least one end of each brace frames into a beam a short distance from a beam-column joint or from another diagonal brace. These short beam segments are called link beams.

Connection A point at which different structural members are joined to each other or to the ground.

Cripple Wall A wood stud wall less than full story height; typically between the floor and foundation wall.

Damage Any unacceptable loadbearing loss or destruction of a structural member.

Deflection Horizontal or vertical movement or displacement (see "Drift").

Diagonal Brace A member of an EBF placed diagonally in the bay of the frame.

Diaphragm A horizontal or nearly horizontal structural element designed to transmit horizontal or earthquake forces to the vertical elements of the seismic resisting system.

Drift Lateral deflection of a building caused by lateral forces.

Dual Frame A structural system with an essentially complete space frame providing support for vertical loads. A moment-resisting frame must resist at least 25 percent of the prescribed seismic forces. The total seismic force resistance is provided by the combination of the moment-resisting frame together with shear walls or braced frames in proportion to their relative rigidities.

Earthquake A sudden motion or vibration in the earth caused by the abrupt release of energy in the earth's lithosphere. The wave motion may range from violent at some locations to imperceptible at others.

Elastic Capable of recovering size and shape after deformation.

Elastic Device A fixed framing restraining device that incorporates an elastic element to reduce the seismic forces transmitted to the structure due to impact from the resilient mounting system.

Epoxy Anchor A type of adhesive anchor using epoxy as the chemical adhesive; see "Adhesive Anchors."

Expansion Anchor An assembly containing a bolt, washer, and nut for connecting to existing concrete elements. The base of the bolt is designed to expand when properly set, wedging the bolt in a predrilled hole.

Fixed A nonyielding or rigid type of restraining device.

Floor Girder A beam that supports floor joists.

High-Temperature Energy Source A fluid, gas, or vapor whose temperature exceeds 220°F (104°C).

Holdown An element or device connected at the ends of the framing of a wall to prevent uplift of the wall.

Horizontal A direction parallel to the ground (sideways).

Inertia Force A force generated by an object as it shakes. The force acts in the opposite direction of the shaking and is related to the weight of the object and its acceleration.

Intensity The apparent effect that an earthquake produces in a structural frame at any given location. In the United States, intensity is frequently measured by the Modified Mercalb Index (MMI).

Intermediate Moment Frame A frame in which members and joints are capable of resisting forces by flexure as well as along the axis of the members.

Inverted Pendulum-Type Structures Framing structures that have a large portion of their mass concentrated near the top and thus have essentially one degree of freedom in horizontal translation. The structures are usually T-shaped with a single column supporting the beams or slab at the top.

Joist Horizontal wood members that support floors or ceilings.

Lateral (Horizontal) Force Resisting System The part of the structural system that has been considered in the design to provide the required resistance to the prescribed seismic forces, i.e., shear walls, braced or rigid frames, floor and roof diaphragms, and foundations.

Lateral Load (Force) Side-to-side force(s) acting on a structure.

Lateral Support Members Secondary members designed to prevent lateral or torsional buckling of beams in an EBF.

Light Framed Wall A framed wall with wood or steel studs, usually a partition or nonloadbearing.

Link Beam The horizontal beam in an EBF which has a length of the clear distance between the diagonal braces or between the diagonal brace and the column face.

Link Beam End Web Stiffeners Vertical web stiffeners placed on the sides of the web at the diagonal brace ends of the link beam.

Link Beam Intermediate Web Stiffener Vertical web stiffeners placed within the link beam.

Link Beam Rotation Angle The angle between the beam outside of the link beam and the link beam occurring at a total story drift of the deflection amplification factor, C_d, times the elastic drift at the prescribed design forces. The rotation angle is permitted to be computed assuming the EBF bay is deformed as a rigid, ideally plastic mechanism.

Load (Dead) The gravity load created by the weight of all permanent structural and nonstructural building components such as walls, floors, roofs, and the operating weight of fixed service equipment.

Load (Live) Moving or movable external loading on a structure. It includes the weight of people, furnishings, equipment, and other things not related to the structure. It does not include wind load, earthquake load, or dead load.

Magnitude A measurement of the relative strength of the earthquake shaking. Magnitude is often reported using the Richter scale.

Mass A quantity or aggregate of matter. It is the property of a body that is a measure of its inertia taken as a measure of the amount of material it contains that causes a body to have weight.

Moment Resisting A structural system with an essentially complete space frame providing support for vertical loads. Seismic force resistance is provided by special, intermediate, or ordinary moment frames capable of resisting the total prescribed forces.

Mudsill/Foundation Sill/Sill Plate The wood member that attaches to the foundation.

Ordinary Moment Frame A frame in which members and joints are capable of resisting forces by flexure as well as along the axis of the members.

P-Delta Effect The secondary effect on shears and moments of frame members due to the action of the vertical loads induced by displacement of the building frame resulting from lateral forces.

Perimeter The outer sides of a building.

Pier A masonry or concrete column used as a beam support structure.

Pier Block A preformed block of concrete used as a footing to support a post.

Post/Column A loadbearing vertical member.

Racking A movement that can distort a structural frame.

Rafter The roof support members.

Resilient Stable Mounting System A framing system incorporating helical springs, air cushions, rubber-in-shear mounts, fiber-in-shear mounts, or other comparable approved systems. The force displacement ratios are equal in the horizontal and vertical directions.

Resonance The amplification of a vibratory movement occurring when the rhythm of an impulse or periodic stimulus coincides with the rhythm of the oscillation (period), for example, when a child on a swing is pushed with the natural frequency of a swing or when an earthquake shakes a building at its own natural frequency.

Restraining Device A device used to limit the vertical or horizontal movement of the mounting system due to earthquake motions.

Richter Scale Named after its creator, the American seismologist Charles R. Richter, a logarithmic scale expressing the magnitude of a seismic disturbance in terms of its dissipated energy.

Seismic Of, subject to, or caused by an earthquake or an earth vibration.

Seismic Activated That part of the structural system that has been considered in the design to provide the required resistance to the seismic forces prescribed in code specs.

Seismic Forces The assumed forces prescribed in the national building codes related to the response of the building's structural framing to earthquake motions to be used in the design of the framing and its components.

Seismic Hazard Any physical phenomenon such as ground shaking or ground failure associated with an earthquake that may produce adverse effects on the structural frame and human activities.

Seismic Risk The probability that social or economic consequences of an earthquake will equal or exceed specified values at a site, at several sites, or in an area during a specified exposure time.

Seismic Strengthening Adding additional bracing, anchoring, or improvement to a structural frame after the original framing construction is completed.

Shear A deformation in which parallel planes slide relative to each other and remain parallel.

Shear Wall A wall, bearing or nonbearing, typically made of wood studs and wood structural panels, built to resist lateral forces from wind or earthquake, from other than its own mass, acting in the direction of the wall.

Sheathing The material covering the surface of the framing of a wall or roof.

Space Frame A structural system composed of interconnected members, other than bearing walls, that is capable of supporting vertical loads and, if so designed, resisting the seismic forces.

Special Moment Frame A frame in which members and joints are capable of resisting forces by flexure as well as along the axis of the members.

Stiffness Resistance to deformation of a structural element or system.

Story Drift Ratio The story drift divided by the story height.

Story Shear The summation of design lateral forces at levels above the story under consideration.

Strength The capability of a material or structural member to resist or withstand applied forces.

Stud The vertical compression members in the walls.

Tension When a connector or wood member resists a pulling force along its axis such as a suspended ceiling or a holdown stud under uplift.

Top Plate The horizontal members that fasten at the top of the studs and support the rafters or joists.

Torque The action or force that tends to produce rotation. In a sense, it is the product of a force and a lever arm as in the action of a wrench twisting a bolt. For example, 50 ft-lb is a force of 50 lb applied at the end of a 1-ft wrench to tighten a bolt.

Uplift Force(s) acting to lift a structure or an element.

Velocity The measure of speed and direction of an object.

Vertical A direction perpendicular to horizontal.

Wall, Bearing A wall providing support for vertical loads; it may be exterior or interior.

Wall, Nonbearing A wall that does not provide support for vertical loads other than its own weight as permitted by the building code; it may be exterior or interior.

Index

About the Author

Jonathan Hutchings is a licensed building contractor and certified construction technologies consultant in California. He also serves on the California State Legal Advisory Board as an Expert Witness and consultant to the insurance industry. He is the author of several books, including *CPM Construction Scheduler's Manual, National Building Codes Handbook,* and *OSHA Quick Guide for Residential Builders and Contractors,* all published by McGraw-Hill.